SOIL MECHANICS IN THE LIGHT OF CRITICAL
AN INTRODUCTION

Soil Mechanics in the Light of Critical State Theories

An introduction

J.A.R. ORTIGAO
Federal University of Rio de Janeiro, Brazil

A.A.BALKEMA/ROTTERDAM/BROOKFIELD/1995

Original text:
Introdução à mecânica dos solos dos estados críticos, Livros Técnicos e Científicos editora
© 1993 J.A.R.Ortigão

Revised and updated edition in English:
©1995 A.A.Balkema, P.O.Box 1675, 3000 BR Rotterdam, Netherlands (Fax: +31.10.4135947)

ISBN 90 5410 194 6 hardbound edition
ISBN 90 5410 195 4 student paper edition

Distributed in USA & Canada by:
A.A.Balkema Publishers, Old Post Road, Brookfield, VT 05036, USA (Fax: 802.276.3837)

Printed in the Netherlands

Contents

Preface

This book evolved from lectures I have given during the last nine years at the School of Engineering of the Federal University of Rio de Janeiro (FURJ). Introducing Critical State Soil Mechanics to undergraduates, I felt the need for a simple text-book to be used in an introductory course on this subject. Eventually it became this book.

The graphical technique of stress paths is used throughout the text. I have chosen, however, the MIT type plot, in lieu of the Cambridge, because they match with Mohr's diagram, which I found easier for an introductory course on the subject.

Mathematical treatment and three-dimensional plots are avoided for the same reason.

In the last chapter, the constitutive equations of the Cambridge models are presented but not derived. They have been employed in a microcomputer program named *Cris*, for teaching purposes, which can be used for simulating triaxial tests on soils. An executable copy of *Cris* is available from the publisher on request.

At FURJ this course is given in one term, after the course of Engineering Geology.

The main references for this course are the books of Atkinson and Bransby and Lambe and Whitman. *SI* units and ISSMFE (The International Society of Soil Mechanics and Foundation Engineering) symbols are employed and recommendations for their application appear in the appendices.

Many colleagues contributed directly or indirectly to this book. It is impossible to acknowledge them all.

Dr Alberto Ortenblad, who has been interested in consolidation since his doctorate in MIT in 1927, revised chapter 7. Just before the publication of this book I received the news that he died in Rio de Janeiro.

Dr Ennio Palmeira, from the University of Brasilia, Dr R. Johnny Fannin from the University of British Columbia, Vancouver, and Dr John Sully from Intevep, Venezuela, read the text and made useful comments.

Many students were very helpful and suggested alterations that made the text clearer. In particular I must thank Tathiane Motta, Alejandro Far, Lucia Alves and Ana Cristina Sieira.

I must also thank Professor R.G. Campanella for having had me as Visiting Associate Professor to the University of British Columbia during 1991, and for the use of UBC facilities for the preparation of this book.

Finally, criticism and suggestions for the next edition are most welcome and should be addressed to the publisher.

JARO

About the author

Beto Ortigao was born in Rio de Janeiro, Brazil, in 1948. He received his first degree in Civil Engineering in 1971 from the Federal University of Rio de Janeiro (FURJ). After working in construction for one year, he returned to FURJ where he obtained a MSc in 1975 and a DSc in 1980, having worked with Professor W.A. Lacerda on the observation of the the behaviour of an instrumented embankment on soft clay.

He became Assistant Professor at FURJ in 1978, and was appointed Associate Professor in 1981. From 1982 to 1984 he visited Britain and worked with Dr K.A. Gallagher on the offshore foundations research programme at the Building Research Establishment, and on offshore foundation design at Fugro UK Ltd.

In 1991 he visited Canada as Visiting Associate Professor to the University of British Columbia to join for one year the In Situ Testing Group headed by Professor R.G. Campanella.

He has been involved in several important engineering projects like: the Juturnaíba and Brumado dams, several embankments on soft foundations, slope stabilization projects including soil nailing, offshore site investigation and foundations at the Campos basin, offshore Rio de Janeiro, pipelines, the foundations of the Sergipe and Manaus harbours, a remote instrumentation system for the slopes of Rio de Janeiro, tunnelling for the Brasília Underground, foundation of bridges and several other structures.

His research interests today encompass in situ testing and instrumentation, soil nailing and field observation of embankments and foundations, on which subjects he has about 50 publications.

He is a member of the ABMS, the Brazilian Society of Soil Mechanics, the ISSMFE, the International Society of Soil Mechanics and Foundation Engineering, the Engineering Club of Rio de Janeiro and a very proud member of the Flamengo Masters Swimming team.

Symbols

Notes

(a)   An apostrophe after a symbol means it refers to effective stresses.

(b)   The subscript f indicates final or failure conditions.

(c)   The subscript ff indicates failure conditions in the failure plane.

(d)   The subscript cr indicates critical state conditions.

(e)   The subscript o indicates initial or in situ conditions.

(f)   The prefix Δ indicates a change.

(g)   Therefix d indicates an infinitesimal value.

Stresses or pressures

p	$p = \dfrac{\sigma_1 + \sigma_2 + \sigma_3}{3}$
p_{atm}	Atmospheric pressure
p'_m	Isotropic overconsolidation stress
q	$q = \sigma_1 - \sigma_3$
s	$s = \dfrac{\sigma_1 - \sigma_3}{2}$
t	$t = \dfrac{\sigma_1 - \sigma_3}{2}$
u	Pore pressure
σ	Normal stress
σ_h	Horizontal normal stress
σ_v	Vertical normal stress
σ_1	Major principal stress
σ_2	Intermediate principal stress
σ_3	Minor principal stress

σ_c	Isotropic confining stress
σ'_{vm}	Vertical effective overconsolidation stress
σ'_g	Intergranular effective stress
σ_{cel}	Cell pressure
σ_{oct}, p	Octahedral normal stress
τ	Shear stress
τ_{oct}	Octahedral shear stress

Relationship between stresses and strains

A	Skempton's pore pressure parameter
B	Skempton's pore pressure parameter $\left(B = \Delta u / \Delta\sigma_3\right)$
CSL	Critical state line
ICL	Isotropic consolidation line
ESP	Effective stress path
K_{cr}	Stress ratio corresponding to the critical state
K_f	K at failure, ie, $\left(\sigma'_3/\sigma'_1\right)_{max}$
K_0	Earth pressure coefficient at rest
OCR	Overconsolidation ratio $OCR = \sigma'_{vm}/\sigma'_v$
SBS	State boundary surface
TSP	Total stress path
α	Henkel's pore pressure parameter
β	Henkel's pore pressure parameter

Displacement and strains

ε	Linear strain
ε_v	Vertical strain
ε_h	Horizontal strain
ε_{vol}	Volumetric strain
ε^e	Elastic strain
ε^p	Plastic strain
γ	Shear strain

Stress strain strength parameters

a'	Intercept of the transformed envelope
c	Cohesion
c_u	Undrained shear strength
c_{ur}	Remoulded undrained shear strength
E	Young's modulus
E'_{oed} or M	Oedometer modulus or Janbu's modulus
G	Shear modulus

K	Bulk modulus
q_c	Tip resistance *CPT*
S_t	Sensitivity $\left(c_u/c_{ur}\right)$
α'	Slope of the transformed envelope
β_1	Slope of the regression line in the Asaoka's method
ϕ	Friction angle
ψ	Dilation angle
η	Slope of the *ESP* $\left(\eta = q/p\right)$
υ	Poisson's ratio

Consolidation

a_v	Compressibility modulus
c_h	Coefficient of consolidation in the horizontal direction
c_v	Coefficient of consolidation
C_c	Coefficient of compressibility, slope of the virgin line
C_s	Coefficient of swelling
CR	Compression ratio $CR = C_c/\left(1 + e_0\right)$
m_v	Volume change modulus
SR	Swelling ratio $SR = C_s/\left(1 + e_0\right)$
U_z	Local degree of consolidation
U	Average degree of consolidation
λ	Slope of the virgin consolidation line in the *e:log p'* diagram
κ	Slope of the swelling line in the *e:log p'* diagram
ρ	Settlement

Physical indices

D_r	Relative density
D_{rc}	Corrected relative density
e	Void ratio
G	Density of grains
n	Porosity
S	Degree of saturation
w	Water content
γ	Unit weight
γ_{sat}	Saturated unit weight
γ_{sub}	Submerged unit weight
γ_d	Dry unit weight
γ_w	Unit weight of water

Other symbols

a	Sample cross section area
A_c	Activity
B	Breadth
C	Hazen's coefficient
D	Diameter
D_{10}	Effective diameter
F	Shape factor
F_p	Seepage force
FS	Factor of safety
f_s	Side friction
g	Acceleration of gravity
GL	Ground level
H	Layer thickness
h_a	Altimetric head
H_d	Drainage path
h_p	Piezometric head
h_t	Total hydraulic total
i	Hydraulic gradient
i_c	Critical gradient
I	Influence factor
I	Rigidity index $= G/c_u$
I_1, I_2, I_3	Stress invariants
J_1, J_2, J_3	Strain invariants
k	Permeability
L	Length
LI	Liquidity index
LL	Liquid limit
NC	Normally consolidated
N_b	Boussinesq's influence factor
n_c	Number of flow channels
n_{eq}	Number of equipotentials
n_{lf}	Number of flow lines
n_s	Number of drops
OC	Overconsolidated
p	Distributed load
PI	Plasticity index
PL	Plastic limit
Q	Point load
Q	Discharge or flow

R	Radius
R_f	Friction ratio $= f_s/q_c$
r	Radius
SL	Shrinkage limit
t	Time
T	Tangential force
T	Torque
T^d	Deviator tensor
T^s	Spherical tensor
T_v	Time factor
V	Volume
V_s	Volume of solids
V_v	Volume of voids
W	Work or energy at deformation
W	Weight
WL	Water level
x, y, z	Coordinate axes
μ	Viscosity
v	Velocity of flow
v_c	Critical velocity
ψ_t	Total hydraulic potential
ψ_c	Kinetic potential
ψ_p	Piezometric potential
ψ_a	Altimetric potential
ψ_k	Thermic potential
ψ_m	Matric potential
$\Re$	Reynolds number
∇	Vector gradient operator

Angles

β	Slope in relation to the horizontal
θ	Slope in relation to the vertical direction
θ_r	Inclination of the failure plane

Test types

CU	Consolidated undrained
CD	Consolidated drained
CIU	Isotropically consolidated undrained
CID	Isotropically consolidated drained
CK_0D	K_0 consolidated drained

CK_0U	K_0 consolidated undrained
CPTU	Piezocone test
FV	Field vane test
SPT	Standard Penetration test
UU	Unconsolidated undrained

Introduction to soil engineering

Introduction

The word soil has to be defined according to its application. For an agronomist, for instance, it is the material that fixes the roots to the ground, and can also be seen as a warehouse from where plants extract nutrients and water. For a mining engineer, the soil overlying an ore pit is simply waste material to be excavated. For a civil engineer, soil is a particle conglomerate from the weathering of rocks, which can easily be excavated and utilized for construction materials or for the support of structures.

As construction or foundation materials, soils are very important for Civil Engineers. In earth dams or in foundations, soils - as concrete or steel bars - are subject to normal and shear stresses and deformation, which ultimately may lead to failure.

Purpose of the book

This book intends to be an introduction to critical state models that enable the calculation of the deformation of a soil element for a given state of stress. These models were developed at the University of Cambridge, England, in the late sixties, and since then have gained widespread use for their simplicity and potential.

The first book on these models was authored by Schofield and Wroth in 1968, being followed by Atkinson and Bransby (1978) and Bolton (1979).

At the Federal University of Rio de Janeiro our interest was drawn only in the 80's (Almeida, 1982; Almeida et al., 1987; Ortigao and Almeida, 1988), but only after 1985 did our MSc courses include these models. The importance of the Cambridge models has led to their introduction at the undergraduate level since 1985, however mathematical treatment and deduction of equations have been avoided.

1

Chapter 1 deals with the origin and formation of soils, index properties, Atterberg limits and grain size analysis. These topics will be presented very briefly. To any reader wishing a more comprehensive coverage I suggest the books by: Hunt (1984), Lambe and Whitman (1979) and Mitchell (1976).

The following two chapters cover total and effective stresses in a soil element and also a review of Continuum Mechanics.

In chapter 4 the effect of loading on the ground is studied. Chapter 5 deals with the steady movement of water in soils. Soil compressibility, settlements and consolidation are presented in chapters 6 and 7. Chapter 8 brings an introduction to soil laboratory tests, which is covered in more detail in chapters 9 to 12 for sands and clays. A few practical applications are studied in chapter 13.

Finally, chapter 14 leads to the calculation of deformation in a soil element through the Cam-clay models. A microcomputer program named *Cris* is used, then, to avoid all the tedious calculations and to allow a quick simulation of soil behaviour through the Cambridge models.

Origin and formation of soils

Soil is a product of the weathering of rocks, i.e., the deterioration that takes places during geological time span.

Weathering processes can be qualified as chemical or mechanical. Chemical weathering is related to the several chemical processes, very common in tropical climates, which alter, dissolve and deposit the rock mineral components, transforming them into soil. Mechanical weathering takes place due to a transporting agent such as water or wind, or to the effect of temperature changes. Frequently many weathering processes occur simultaneously.

Soils that remain at the site of their formation are called *residual*; those that suffer the action of transporting agent are called *transported* or *sedimentary*, the latter is applied to water-transported soils.

A typical profile of a residual soil is presented in figure 1.1. At great depth the bedrock keeps its original characteristics and remain sound. As it gets closer to the surface the weathering increases. Overlying bedrock, weathered and fractured rock occurs, which allows seepage of water through its joints and faults. Just above, there is a layer of *young residual soil* or *saprolite*, still keeping rock structures such as bedding and shear planes.

The layer above the saprolite has been subjected to great weathering and has lost all rock structures, having a major percentage of clay. Therefore, it is called *mature residual soil*.

Figure 1.2 presents a borehole log of a residual soil, extending down to the bedrock.

Water transported soils can be classified as *alluvium, lacustrine* and *marine* soils, as they are formed in river, lake or sea water environment. An example of the latter is shown in figure 1.3. Such deposits are formed in bays or rivers' estu-

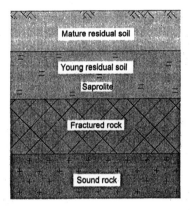

Figure 1.1. Typical soil profile in residual soil

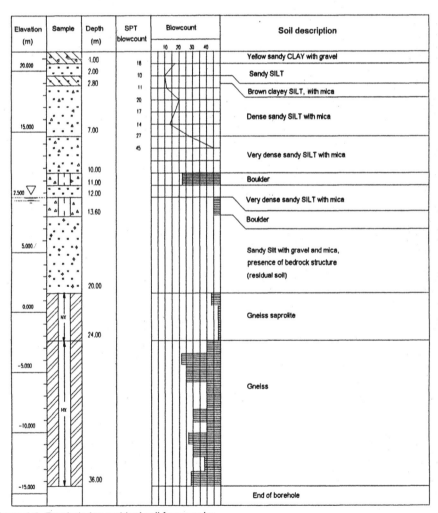

Figure 1.2. Borehole log residual soil from gneiss

Figure 1.3. Typical soil profile in soft soil

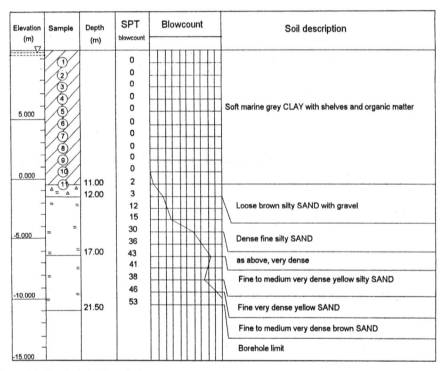

Figure 1.4. Borehole log in soft clay

aries due to the sedimentation of transported soil. A typical borehole log is presented in figure 1.4 consisting of a thick layer of clay overlying several layers of sand.

Figure 1.5 shows a type of soil known as *colluvium* or *talus*, encountered at the base of steep mountains. This is a common material in Rio de Janeiro and

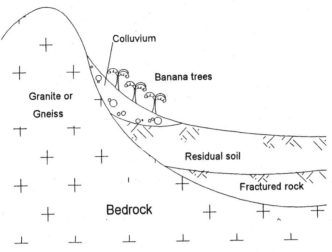

Figure 1.5. Colluvial soil or talus

Hong Kong where granite or gneiss rocks are weathered and transported to the base of the hills, forming a deposit of soft mass, containing many boulders and blocks of rock. During summer rainstorms they become saturated and cause landslides.

In a tropical environment one can spot a talus through the type of vegetation growing on it. Banana trees have a special preference for these soils due to their high water content and low resistance to the spreading of the tree roots.

Till is a post-glaciation sedimentary soil deposited as a result of the retreat and melting of glaciers. Its formation results in a stiff conglomerate of a wide range of grain sizes from silts to gravels.

Soil characterization

Some soil properties that can be determined through simple soil tests are specially useful for soil characterization. These simple tests are: grain size distribution, unit weight, water content, void ratio and Atterberg limits, which will be described as follows:

Grain size distribution

The analysis of grain size distribution is aimed at obtaining a plot as shown in figure 1.6. It relates particle diameter to percentage of soil passing or being retained in the sieve.

In coarse materials, like sands and gravels, this is accomplished by sieving a soil sample in a standard series of sieves and measuring the mass of the retained soil particles in each one. As an example, consider a series of sieves having different mesh diameters shown in table 1.1.

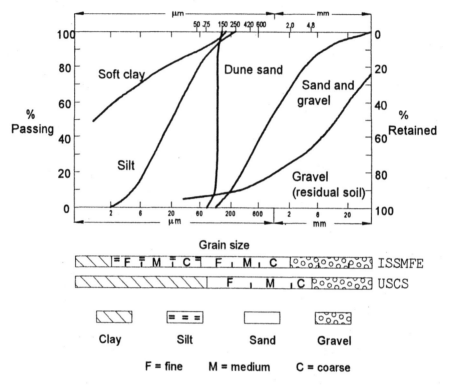

Figure 1.6. Grain size distribution in different soil types (Vargas, 1977)

Table 1.1. Standard series of sieves

Mesh diameter
4.8 mm
2.0 mm
600 μm
420 μm
250 μm
75 μm

After sieving a soil sample, the mass of the retained material in each sieve is measured, and the data plotted as shown in figure 1.6, in which the horizontal axis is logarithmic. This test is standardized in many countries.

Very fine soils, having particle diameter smaller than 75 μm, are treated separately by a sedimentation analysis, the details of which can be found elsewhere (e.g., Lambe and Whitman, 1979, Hunt, 1984).

The results of a grain size distribution are interpreted through the comparison with standard grain size scale, as shown in figure 1.6. The first is the International Scale, recommended by the ISSMFE - the International Society for Soil Mechanics and Foundation Engineering. It is the simplest to memorize because it is based on numbers 2 and 6, as shown in table 1.2. It was recommended long ago in order to unify several different classification systems, having been proposed in a Soil Science conference in 1927 (Means and Parcher, 1965), and already been adopted as standard in most countries.

In the USA, the preferred system for soil classification is still the USCS - Unified Soil Classification System, proposed in 1948 by the late Professor Casagrande (1948) and summarized in table 1.3. Details on this system are described elsewhere in laboratory and site investigation manuals, e.g., Hunt (1984), Holtz and Krizek (1984).

The USCS classification system divides soils into gravels (*G*), sands (*S*), silts

Table 1.2. International Scale

Soil description	Diameter
Clays	< 2 μm
Silt	2 μm to 60 μm
Fine sand	60 μm to 200 μm
Medium sand	200 μm to 600 μm
Coarse sand	600 μm to 2 mm
Gravel	> 2 mm

Table 1.3. USCS Soil Classification System

Symbol	Soil type
GW	Well graded gravels, gravel sand mixtures, little or no fines
GP	Poorly graded gravels, gravel sand mixtures, little or no fines
GM	Silty gravels, gravel sand silts mixtures
GC	Clayey gravels, gravel sand clay mixtures
SW	Well graded sands, gravelly sands, little or no fines
SP	Poorly graded sands, gravelly sands, little or no fines
SM	Silty sands, sand silt mixtures
SC	Clayey sands and clay mixtures .
ML	Inorganic silts and very fine sands
CL	Inorganic clays of low to medium plasticity
OL	Organic silts and organic silty clays of low plasticity
MH	Inorganic silts, micaceous fine sandy silty soils
CH	Inorganic clays of high plasticity
OH	Organic clays of medium to high plasticity, organic silts
Pt	Peat or other highly organic soils

(*M*, which stands for the Swedish word *mo* for silt), clays (*C*), organic soils (*O*) and peats (*Pt*). In coarse materials, such as gravels and sands, a second letter *W* or *P* (Well graded or Poorly graded) can be added for describing its size distribution, as commented in the following paragraph. In fine soils, such as silts and clays, the letters *L* and *H*, standing for low and high, gives an indication of the liquid limit of the soil, whether it is higher or lower than 50%.

Figure 1.6 presents an example of the grain size distribution for different soils. The gravels, on the right side of the figure, show a gentle slope curve and are called *well-graded* material. On the other hand, the dune sands in the middle of the figure, indicate predominance of one single diameter, and therefore are called *poorly graded*. This is due to the fact this sand has been transported by the wind.

Sand grains can be classified according to their shape (figure 1.7) as *angular*, *subrounded* or *rounded*. The last one is characteristic of river sand.

Physical indices

Consider a soil element (figure 1.8) containing solid particles, water and air, where:

V = total volume;
V_v = volume of voids;
V_a = volume of air;
V_w = volume of water;
V_s = volume of solids;

and

W = total weight;
W_w = weight of water;
W_s = weight of solid particles;
γ_w = unit weight of water, taken as 10 kN/m³.

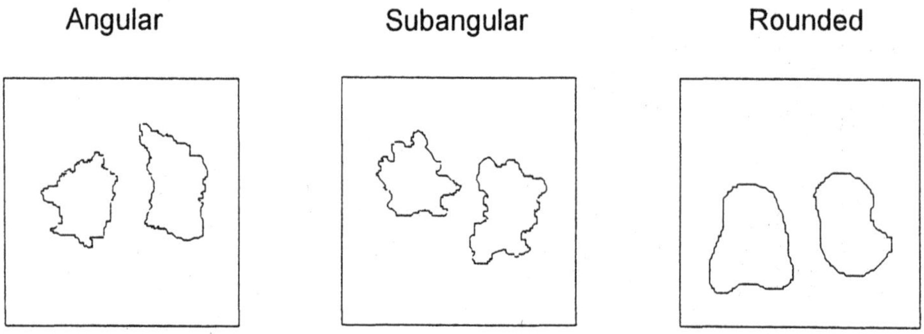

Angular	Subangular	Rounded

Figure 1.7. Sand grain shapes

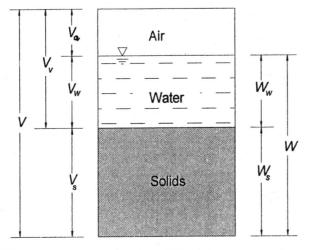

Figure 1.8. Weight and volume of an unsaturated soil element

Table 1.4. Physical indices

Void ratio	e	$e = \dfrac{V_v}{V_s}$
Porosity	n	$n = \dfrac{V_v}{V}$
Degree of saturation	S	$S = \dfrac{V_w}{V_v}$
Water content	w	$w = \dfrac{W_w}{W_s}$
Total unit weight	γ	$\gamma = \dfrac{W}{V}$
Saturated unit weight	γ_{sat}	as above, for S = 100 %
Submerged unit weight	γ_{sub} or γ'	$\gamma_{sub} = \gamma_{sat} - \gamma_w$
Dry unit weight	γ_d	$\gamma_d = \dfrac{W_s}{V}$
Specific gravity	G_s	$G_s = \dfrac{\gamma_d}{\gamma_w}$

Table 1.4 summarizes the physical indices most used in soil mechanics and presents their definition. Some of them can be related. Useful relationships between them are shown in table 1.5.

Comments on the physical indices:

The *void ratio e* is used to represent the volume state of the soil. Volumetric

Table 1.5. Relationship between physical indices

$$n = \frac{e}{1+e}$$

$$e = \frac{n}{1-n}$$

$$G_s w = Se$$

$$\gamma = \frac{G_s(1+w)}{1+e}\gamma_w$$

$$\gamma_d = \frac{\gamma}{1+w}$$

$$\gamma_{sat} = \frac{G_s+e}{1+e}\gamma_w$$

strains, as will be shown later in chapter 6, are proportional to a void ratio variation Δe. The greater the void ratio, the greater the volumetric deformation the material will present when loaded in certain conditions.

The *degree of saturation S* is equal to 100% in fully saturated materials, nil for dry ones.

The *water content w* has little significance in sands, but in clays it can be related to the volume change and strength behaviour.

The *total unit weight* γ or γ_t allows the calculation of stresses in the soil mass due to overburden material, as will be discussed in detail in chapter 3. Most clays present a value of γ ranging from 13 to 20 kN/m³. In sands a typical value will lie in the 17 to 20 kN/m³ range.

The *saturated unit weight* γ_{sat} refers to γ value for a fully saturated material.

The *submerged* or *buoyant unit weight* γ_{sub} or γ' is used to calculate average intergranular stresses, or effective stresses. This will be discussed in detail in chapter 3.

The *dry unit weight* γ_d refers to the unit weight for a soil without moisture.

The *specific gravity* G_s is a relationship between the unit weight of solids of a soil and the unit weight of water, and is a pure number. For most soils the specific gravity G_s lies in the 2.7 ± 0.1 range. For quartz it is $G_s = 2.65$. However, for soils containing iron ore, like the hematite and magnetite, the G_s value can be as high as 5.

Exercise 1.1

A saturated soil sample with a volume of 560 cm³ and mass of 850 g was dried in an oven during 24 h at 105°C, and the mass decreased to 403 g. Let $G_s = 2.7$. Obtain w, e, and γ.

Solution

(a) w

$$w = \frac{W_w}{W_s} = \frac{M_w}{M_s} = \frac{850 - 403}{403} = 1.11 = 111\%$$

(b) e; from equation $G\,w = S\,e$, comes:

$$e = \frac{G_s w}{S} = \frac{2.7 \times 1.11}{1} = 3.00$$

(c) γ

$$\gamma = \frac{W}{V} = \frac{850 \times 10^{-3} \times 9.81 \times 10^{-3}}{560 \times (0.01)^3} = 14.9 \, \text{kN/m3}$$

Exercise 1.2

A sample of an alluvial clayey sand from São Paulo (figure 1.17) gave the following data: $G_s = 2.72$, $e = 0.75$ and $S = 50\%$. Obtain w, γ, γ_{sat}, γ_{sub} and γ_d.

Solution

(a) w

$$w = \frac{Se}{G_s} = \frac{50 \times 0.75}{2.72} = 14\%$$

(b) γ; from table 1.5 the following equation is used:

$$\gamma = \frac{G_s(1+w)}{1+e} \gamma_s = \frac{2.72(1+0.14)}{1+0.75} 10 = 17.7 \;\text{kN/m}^3$$

(c) γ_{sat}

$$\gamma_{sat} = \frac{G_s + e}{1+e} \gamma_w = \frac{2.72 \times 0.75}{1+0.75} 10 = 19.8 \;\text{kN/m}^3$$

(d) γ_{sub}

$$\gamma_{sub} = \gamma_{sat} - \gamma_w = 19.8 - 10 = 9.8 \;\text{kN/m}^3$$

(e) γ_d

$$\gamma_d = \frac{\gamma}{1+w} = \frac{17.7}{1+14/100} = 15.5 \;\text{kN/m}^3$$

Relative density of sands

The relative density of a sand is given by

$$D_r = \frac{e_{max} - e}{e_{max} - e_{min}} \tag{1.1}$$

where: e_{max} is the maximum void ratio when the sand is in the loosest state; e_{min}, the minimum, when it is in the densest state, and e the current void ratio.

D_r is commonly expressed as a percentage. When the sand presents D_r less than 30% it is regarded to be loose, between this value and 70%, medium dense, and above 70%, dense.

The relative density has considerable practical importance, being used for quality control of embankments that employ a sandy material. It is generally required that $D_r > 70\%$ after compaction.

Atterberg limits

In 1911 the Swedish agronomist Atterberg published a paper (*Über der Physicalische Bodenuntersuchung und über die Plasticität der Tone,* Internationale Mitteilungen Bodenkunde, vol. 1, pp 10-43) classifying the water content values that a clay can present in limits corresponding to the apparent state of the material, as shown in figure 1.9.

The following limits were defined as follows: *shrinkage limit SL, plastic limit PL* and *liquid limit LL,* corresponding to the transition between the solid state, when volume change ceases, to the plastic state, when volume varies with the water content, and the liquid state.

Atterberg also suggests that the difference between the liquid and plastic limits, named *plasticity index PI,* could be used for soil classification:

$$PI = LL - PL \qquad (1.2)$$

This subject was studied by Casagrande many years later, who designed the laboratory equipment shown in figure 1.10 for the liquid limit test, conducted as follows:

A soil sample is mixed in a container and its water content w is determined. It

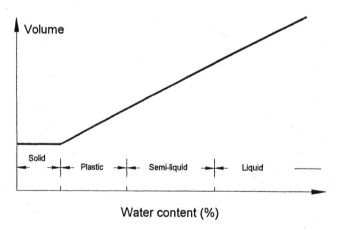

Figure 1.9. Volume water content relationship

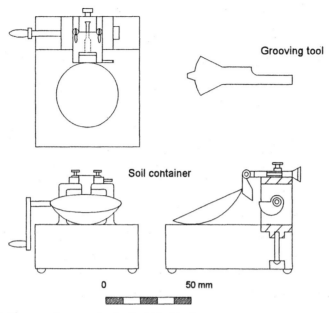

Figure 1.10. Casagrande apparatus for liquid limit test

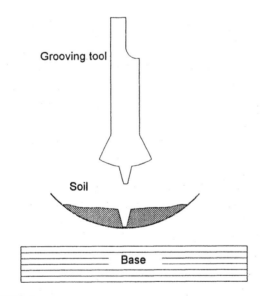

Figure 1.11. Liquid limit test

is, then, placed in the apparatus container and a longitudinal groove, as indicated in figure 1.11, is made in the sample. The container is raised to a standard height and dropped on the equipment base. This is repeated several times until the base of the groove in the soil sample closes. The number of blows is then recorded.

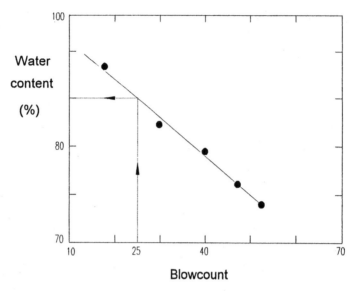

Figure 1.12. Blowcount versus water content relationship obtained in liquid limit test

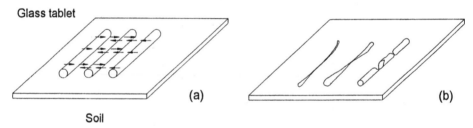

Figure 1.13. Plasticity limit test

Water is added to the sample and the test is repeated for several water contents. The results are plotted as in figure 1.12. The liquid limit *LL* is defined as the water content corresponding to 25 blows.

The plastic limit test aims at determining the water content that starts fracturing a cylindrical soil sample, of approximately 3 mm in diameter, as shown in figure 1.13. The sample is rolled by hand in a see-saw movement. The plastic limit is the water content in which the sample starts to crumble.

Activity

The colloidal activity of clays was studied by Skempton (1953) who defined activity A_c as:

Table 1.6. Activity of clays.

A_c	Activity
< 0.75	inactive
0.75 - 1.25	normal
>1.25	active

$$A_c = \frac{PI}{\text{clay fraction}} \qquad (1.3)$$

The clay fraction corresponds to the percentage of material under 2 μm in diameter.

Skempton classified clays according to table 1.6.

The clay activity value is an indication of volume change susceptibility. Active clays may present more volume change than inactive clays, when subjected to the same loading.

Liquidity index

The *liquidity index* of a clay is defined by:

$$LI = \frac{w - PL}{PI} \qquad (1.4)$$

where: w is the water content of the sample.

When a clay sample presents $w = LL$, $LI = 1$; when $w > LL$, then $LI > 1$.

This property is used for soil classification and is generally an indication of the sensitivity of the clay, a characteristic that will be discussed later on in chapter 12. In general, clays with LI above 1 are usually regarded as sensitive soils.

Exercise 1.3

The Rio de Janeiro clay (figure 1.15) presents the following average data: $LL = 120\%$, $PL = 40\%$, $w = 150\%$. The clay content, i.e., the percentage of the material under 2 μm in diameter is 55%. Obtain the plasticity index, the activity and the liquidity index.

Solution

(a) Plasticity index:

$$PI = LL - PL = 120 - 40 = 80\%$$

(b) Activity:

$$A_c = \frac{PI}{\% < 2\mu m} = \frac{80}{55} = 1.45$$

(c) Liquidity index:

$$LI = \frac{w - PL}{PI} = \frac{150 - 40}{80} = 1.4$$

Soil profiles

Soil parameters and indices studied in this chapter can be plotted with depth for a given soil profile, and this can be very useful for soil identification and characterization. A few examples can be seen in figures 1.14 to 1.17 for soils of different geological origin.

Soft marine and deltaic clays are common deposits in many countries, e.g., the San Francisco bay mud in the USA, the Osaka bay clay in Japan and the Fraser river deposits in Vancouver, Canada. Figure 1.14 presents typical data on the Rio de Janeiro marine clay that covers Guanabara bay. Atterberg limits, in situ void ratios, total unit weights have been plotted, as well as the undrained shear strength c_u that will only be discussed later on in chapter 12. Typical average values for this clay are: $PL = 40\%$, $LL = 120\%$, hence $PI = 80\%$. A plasticity index of 80 % is a very large value. The water content is greater than LL, therefore the liquidity index is higher than 1. The in situ void ratio e is typically 4 on the top of the clay layer, decreasing to 3 at the bottom. The unit weight γ varies between 13 to 14 kN/m³ .

The geotechnical properties of another marine clay from the northeastern state of Sergipe, Brazil, are shown in figure 1.15. The clay lies between sand layers. LL is around 80%, PL around 40%, therefore, $PI = 40\%$. Unit weight,

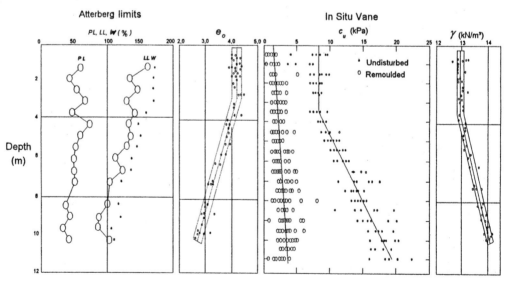

Figure 1.14. Geotechnical properties of Rio de Janeiro soft clay

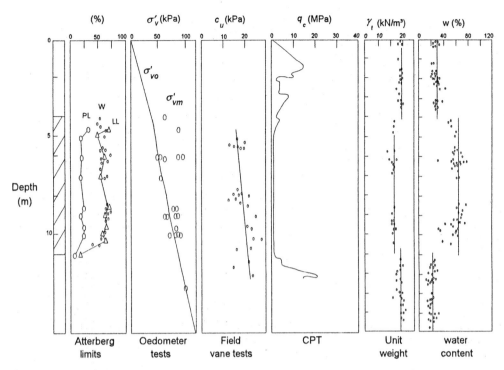

Figure 1.15. Geotechnical properties of soft clay deposit from Sergipe, Brazil (Ortigao and Sayao, 1994)

water content and particle diameter data are shown in the figure, as well as the undrained strength c_u, and q, the cone point resistance, which will be discussed in chapter 7.

A special kind of sand, known as *calcareous* or *carbonated* sand has been found in many countries and has been the cause of numerous foundation problems. It has grains of calcareous material that are easily crushed, giving rise to a high compressibility, when compared to an onshore sand deposit, made of hard grains of silica or quartz. Calcareous sands are formed by organic debris and are mainly found in warm sea waters between parallels ±30°, as in the North West Shelf of Western Australia and in the Bass Strait in South East Australia, as well as in several sites in the Middle East.

The offshore soil deposit shown in figure 1.16 is found about a hundred kilometers from the Brazilian coast, at a water depth around 100 m. These data refers to the Carapeba site, at the Campos basin, Rio de Janeiro, where several oil exploration structures have been built since 1980 (Ortigao et al., 1986). Grain crushability in calcareous sands leads to important geotechnical foundation problems discussed elsewhere (e.g. Ortigao et al., 1985). The identification of these sands can be made through $CaCO_3$ content, as indicated in the figure for depths from 20 to 120 m.

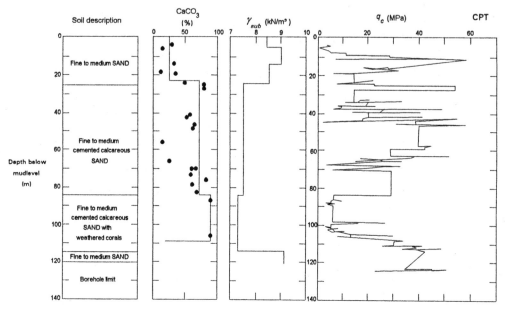

Figure 1.16. Geotechnical properties of an offshore calcareous sand from the Campos basin, Brazil (Ortigao et al., 1986)

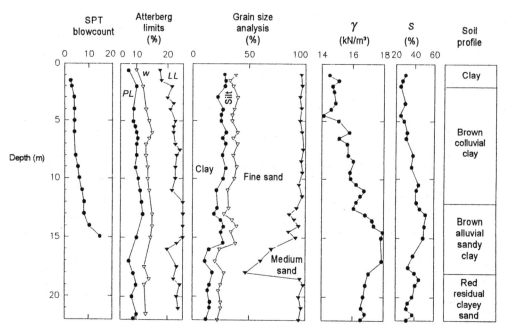

Figure 1.17. Geotechnical properties from a compressible soil from São Paulo, Brazil (Ferreira and Monteiro, 1985)

Finally, a summary of the geotechnical properties of an unsaturated colluvial soil overlying a residual soil of sandstone, São Paulo, is shown in figure 1.17 (Ferreira and Monteiro, 1985). Atterberg limits are approximately: $LL = 25\%$, $PL = 5\%$. Due to the very dry environment in which it has been formed, both the water content w and the degree of saturation S are very low, in the order of 10 and 20%, respectively. This soil presents an important feature: great volume compression when wet, being known for this reason as *collapsible* soil, which brings important consequences for the foundations of buildings. This subject will be reviewed in chapter 6.

Proposed exercises

1.1. A saturated soil sample having a volume of 300 cm³ presents a mass of 423 g. After being fully dried in an oven at 105°C, its mass decreased to 320 g. Considering $G_s = 2.65$, obtain the water content w, the initial void ratio and the total, dry and submerged unit weight.

1.2. Repeat previous exercise considering that soil sample is obtained from an iron ore mining region and has a high percentage of hematite, in which $G_s = 5$.

1.3. A rockfill is being constructed with granite rock blocks ($G_s = 2.7$), presenting a void ratio of 0.5 after placement. Evaluate the apparent total, dry and submerged unit weight.

1.4. Using the data shown in figure 12.14 for the Rio de Janeiro clay, obtain plots of LI and PI against depth.

1.5. Considering that the Rio de Janeiro clay presents 55% of particles with diameter under 2 μm, obtain a plot of activity of this clay against depth. Classify the results according to Skempton.

1.6. A sand embankment is to be constructed and design specifications require a minimum relative density of 70%. If $e_{min} = 0.565$ and $e_{max} = 0.878$ for the sand, what should the void ratio be after placement ?

1.7. For the upper sand layer shown in figure 1.15, obtain the dry unit weight and the void ratio, taking $G_s = 2.69$.

Chapter 2

Stresses and deformation in soils

Introduction

This chapter presents a summary of continuum mechanics with emphasis on Soil Mechanics applications. Only the main concepts and important equations, but no formulae deduction, will be presented. A comprehensive coverage of this subject is out of the scope of this book but can be found in more detail in the following references Timoshenko and Goodier (1951), Poulos and Davis (1974) or Harr (1966).

The concept of stress

Consider a body in equilibrium under the action of external forces (figure 2.1). Plane A divides it in two parts that are kept in equilibrium by the action of external and internal forces. The internal forces act on the cross section produced by plane A. Consider the elementary area dA and n its normal line. The elementary internal forces acting on this tiny area have dF as a resultant, which can be split into the normal and tangential components dN and dT. Therefore, one can define the normal and shear stresses referred to plane A as:

Normal stress: $\quad \sigma_n = \lim_{dA \to 0} \dfrac{dN}{dA}$

Shear stress: $\quad \tau = \lim_{dA \to 0} \dfrac{dT}{dA}$

The elementary tangential force dT can also be split into its components dT_x and dT_y, relative to the coordinated axes x and y (figure 2.2). Therefore, one can define the following shear stresses:

$\tau_x = \lim_{dA \to 0} \dfrac{dT_x}{dA}$

20

$$\tau_y = \lim_{dA \to 0} \frac{dTy}{dA}$$

The ultimate purpose of this study is to find out the state of stress (and deformation) of any point in a soil mass. This implies that the normal and shear stresses (and deformation), relative to any plane that intercepts the chosen point, are known. It will be seen, however, that provided we know the stresses relative to three orthogonal planes, the state of stress can be found out.

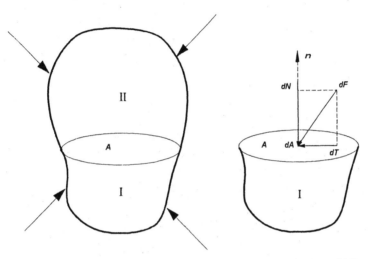

Figure 2.1. (a) Equilibrium conditions in the continuum under external loading; (b) Decomposing internal forces acting at an elementary area *dA*

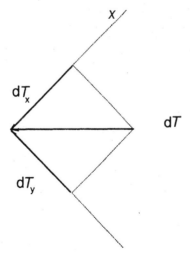

Figure 2.2. Decomposing the elementary force *dT*

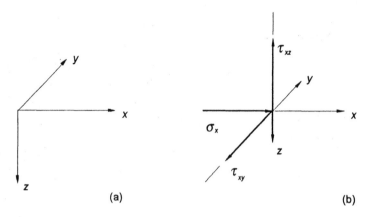

Figure 2.3. (a) 3 orthogonal planes; (b) Decomposing normal and shear stresses according to three orthogonal planes

Consider a single point in a body in equilibrium and three orthogonal planes containing the point defined by the coordinated axes x, y and z (figure 2.3). It is possible to define normal stresses relative to these 3 planes: σ_x, σ_y and σ_z, and two shear stresses for any normal stress. This results in 3 normal and 6 shear stresses. Arranged in matrix form, these stress components constitute what is known as the *stress tensor* with 9 components all together.

$$|\sigma| = \begin{vmatrix} \sigma_x & \tau_{xy} & \tau_{xz} \\ \tau_{yx} & \sigma_y & \tau_{yz} \\ \tau_{zx} & \tau_{zy} & \sigma_z \end{vmatrix}$$

Conditions for equilibrium

Considering the equilibrium conditions in the vicinity of a point, it is possible to conclude that:

$$\tau_{xy} = \tau_{yx}$$
$$\tau_{zx} = \tau_{xz}$$
$$\tau_{zy} = \tau_{yz}$$

Consequently, the nine components of the stress tensor are reduced to only six independent terms.

Stresses relative to a plane

Take point A in a continuum. Consider that one knows the normal and shear stresses at a point A according to three orthogonal planes that contain A. Given new plane N, it can be prooved that one can find out the the stresses at A according to plane N.

Consider plane N, defined by the cosines of the directions of its normal n: $\cos(n,x)$, $\cos(n,y)$, and $\cos(n,z)$, each one being the cosine of the angle between the normal n and the axes x, y or z.

Let p_n be the resultant of the stresses at a point relative to plane N and, p_{nx}, p_{ny} and p_{nz} the p_n components according to x, y and z axes. It is possible to show that this resultant can be obtained through the following matrix equation:

$$\begin{vmatrix} p_{nx} \\ p_{ny} \\ p_{nz} \end{vmatrix} = \begin{vmatrix} \sigma_x & \tau_{xy} & \tau_{xz} \\ \tau_{yx} & \sigma_y & \tau_{yz} \\ \tau_{zx} & \tau_{zy} & \sigma_z \end{vmatrix} \begin{vmatrix} \cos(n,x) \\ \cos(n,y) \\ \cos(n,z) \end{vmatrix} \tag{2.1}$$

It can be concluded, therefore, that if one knows the normal and shear stresses that act at a point relative to three orthogonal planes, the state of stress is known. In other words, if the stress tensor is known, the state of stress is also known. In addition, the stress tensor is the base in the vector space $\Re^3$.

Transformation of coordinates

Previous conclusions allow to verify that equation 2.1 can be used to change all reference planes, i.e., to choose a new coordinate system of axes x, y and z and relate the stress tensor to this new system. Starting from equation 2.1 it is possible to demonstrate that:

$$|\sigma_1| = |A| \times |\sigma| \times |A|^T \tag{2.2}$$

where: $|\sigma_1|$ is the stress tensor referred to the new coordinate system x, y and z; $|A|$ is the matrix of cosines, below:

$$|A| = \begin{vmatrix} \cos(x_1,x) & \cos(x_1,y) & \cos(x_1,z) \\ \cos(y_1,x) & \cos(y_1,y) & \cos(y_1,z) \\ \cos(z_1,x) & \cos(z_1,y) & \cos(z_1,z) \end{vmatrix}$$

$|A|^T$ is the transpose matrix of $|A|$

Principal stresses

It is possible to demonstrate that for any point of the continuum, there will be a system of coordinate axes x^*, y^*, z^* for which the shear stresses are nil and the normal stresses present maximum and minimum values.

This demonstration starts from equation 2.2. All the shear stresses appearing in the stress tensor $|\sigma_1|$ are made equal to zero. The resulting equation is then solved to find the normal stresses. This leads to an equation of the third degree known as *characteristic equation:*

$$\sigma_i^3 - I_1\sigma_i^2 + I_2\sigma_i - I_3 = 0 \tag{2.3}$$

The roots of this equation are the *principal stresses* σ_1, σ_2 and σ_3. It is conventionally adopted that $\sigma_1 > \sigma_2 > \sigma_3$, and these stresses are called respectively *major*, *intermediate* and *minor* principal stresses.

The coefficients of the characteristic equation are:

$$I_1 = \sigma_x + \sigma_y + \sigma_z \tag{2.4}$$

$$I_2 = \sigma_x\sigma_y + \sigma_y\sigma_z + \sigma_z\sigma_x - \tau_{xy}^2 - \tau_{xz}^2 - \tau_{zy}^2 \tag{2.5}$$

$$I_3 = \sigma_x\sigma_y\sigma_z - \sigma_x\tau_{zy}^2 - \sigma_y\tau_{xz}^2 - \sigma_z\tau_{xy}^2 - 2\tau_{xy}\tau_{yz}\tau_{zx} \tag{2.6}$$

The terms I_1, I_2 and I_3 are independent of the coordinate system, as can be observed through equations 2.4 to 2.6. They are called as *stress invariants*.
Equations 2.4 to 2.6 can be simplified if one chooses the appropriate coordinate system that corresponds to the principal stresses. In this case, all shear stresses turn to nil, and one obtains:

$$I_1 = \sigma_1 + \sigma_2 + \sigma_3 \tag{2.7}$$

$$I_2 = \sigma_1\sigma_2 + \sigma_2\sigma_3 + \sigma_1\sigma_3 \tag{2.8}$$

$$I_3 = \sigma_1 \ \sigma_2 \ \sigma_3 \tag{2.9}$$

Octahedral stresses

Dealing with stress-strain behaviour of materials, it is frequently necessary to use a mean value of the normal stresses defined as the *octahedral normal stress*, given by:

$$\sigma_{oct} = \frac{1}{3}(\sigma_x + \sigma_y + \sigma_z) \tag{2.10}$$

This value is independent of the coordinate system, since:

$$\sigma_{oct} = \frac{1}{3}I_1 \tag{2.11}$$

Searching the direction of planes where σ_{oct} acts, one finds that these planes form an angle of $\cos^{-1}\sqrt{1/3}$ with the direction of principal stresses. They intersect each other forming a figure of an octahedron, after which the stress was named.

The octahedral shear stress is given by the following equation as function of the principal stresses:

$$\tau_{oct} = \sqrt{\frac{1}{3}\left[(\sigma_1 - \sigma_2)^2 + (\sigma_2 - \sigma_3)^2 + (\sigma_3 - \sigma_1)^2\right]} \qquad (2.12)$$

It can also be observed that τ_{oct} can be obtained from the stress invariants through the following equation:

$$\tau_{oct} = \pm \sqrt{\frac{2}{9}\left[I_1^2 - 3I_2^2\right]} \qquad (2.13)$$

Another important notation for the normal octahedral stress is p, and will be in used from chapter 3 onwards for the stress path diagrams. Remember that:

$$p = \sigma_{oct}$$

Two-dimensional space

In many practical applications a three dimensional problem can be simplified to two dimensions only. This can lead to an important simplification in the mathematical treatment of engineering problems. A frequent case, when dealing with stresses and deformations, is the *plane strain* case.

Consider the example in figure 2.4a of an earth dam with a much bigger length (along the x axis) than the dimensions across. This implies that all the deformation during construction and operation of the dam will be concentrated in the cross sectional planes parallel to yz. The minor and major principal stresses σ_1 and σ_3 occur in these cross-sections planes, but the intermediate principal stress σ_2 is parallel to x. It can be demonstrated that σ_2 is not an independent value, but a function of the principal stresses σ_1 and σ_3. Therefore, the problem

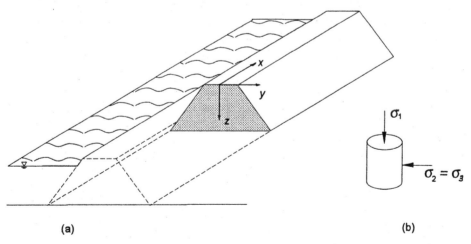

| (a) | (b) |

Figure 2.4. (a) Plane strain conditions in an earth dam; (b) Axi-symmetrical condition in a soil specimen

can be treated in two dimensions only. This simplification is very important in many applications.

Other examples of engineering problems that can be dealt with in plane strain are: very long retaining walls, roads, and many others.

Axi-symmetrical condition

Another kind of problems presenting an axial symmetry can also be treated in two dimensions. Examples of axial symmetry are: a soil specimen used in laboratory test (figure 2.4b), a tubular pile foundation, a tower, as many other structures. The problem can be reduced to an axi-symmetrical condition with great simplification since: $\sigma_2 = \sigma_3$.

Mohr's circle

Mohr's circle is a graphical representation of the state of stress of a single point of the continuum. It is a very useful technique for all stress and deformation problems and will be used many times throughout this book. Therefore, it is strongly recommended that the reader gets familiarized with the Mohr's circle before stepping ahead.

Figure 2.5a presents a two-dimensional stress condition in plane yz in which the normal stresses σ_y, σ_z and the shear stress τ_{yz} are known. The corresponding Mohr's circle (figure 2.5c) is obtained in the following way:

1. Plot a Cartesian coordinate system (σ, τ) in which the abscissae are the normal stresses σ and the ordinates, the shear stresses τ.
2. Choose a plane and their related normal and shear stresses for plotting, e.g., the vertical plane xy, where σ_z and τ_{yz} act.
3. Work out the sign for the shear stress τ_{yz}, according to the convention shown on the right side of figure 2.5a. Imagine a point outside the plane where the shear stress acts, then, workout the direction of the rotation of the shear stress around this imaginary point. Clockwise rotation means *positive*, anticlockwise means *negative*.
4. In soil mechanics, compression is positive, tension is negative. This is just the contrary of what is used in concrete, for instance, and in many text-books of strength of materials.
5. Plot the point having (σ_y, τ_{yz}) coordinates, in which τ_{yz} positive.
6. Plot the point having coordinates (σ_z, τ_{zy}), in which τ_{zy} is negative as shown in figure 2.5a, since this stress rotates anticlockwise.
7. The line linking (σ_y, τ_{yz}) to (σ_z, τ_{zy}) intercepts the abscissae in a point corresponding to the centre of Mohr's circle, which can now be plotted.
8. Mohr's circle intercepts the abscissae at points corresponding to the principal stresses, where shear stresses are zero, as indicated in figure 2.5c.

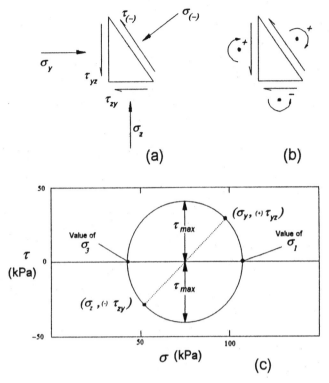

Figure 2.5. (a) Two dimensional stress condition; (b) sign convention for shear stress τ; (c) Mohr's circle

9. Other important points on Mohr's circle are those corresponding to the maximum and minimum shear stresses, also shown in figure 2.5c. Note that τ_{max} is equal to τ_{min} in modulus.

The pole

The pole of the Mohr's circle is an auxiliary technique that enables graphical determination of normal and shear stresses for a given plane, or vice-versa. This is a very useful graphical technique shown in figure 2.6. For a given Mohr's circle and for a given direction θ of inclination of a plane, the pole technique yields the stresses σ_θ and τ_θ. It can be applied as follows:

The first step is to work out the position of the pole:

1. Choose any point on the Mohr's circle in which the direction of the corresponding plane is known (point *1* in figure 2.6).
2. From point *1*, draw a line parallel to the plane direction until it crosses the circle again.
3. This interception is point *2* where lies the pole.

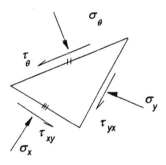

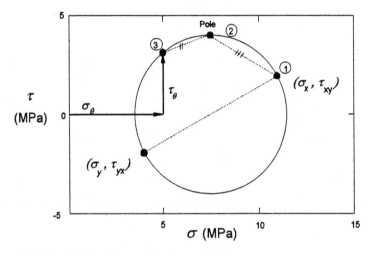

Figure 2.6. Determination of σ_θ and τ_θ through Mohr's circle utilizing the pole construction

Thereafter, obtaining stresses for any given plane is straightforward:

4. Starting at the pole, draw a parallel line to the direction of the given plane.

5. This line crosses Mohr's circle at point 3, corresponding to the stresses σ_θ and τ_θ, as required.

Exercise 2.1

In the Mohr's circle shown in figure 2.7, point A corresponds to a vertical plane. Obtain: (1) the pole; (2) the stresses acting in the horizontal plane; (3) values of σ_1 and σ_3 and the direction of planes where they act; (4) τ_{max} and τ_{min} and the direction of the planes where they act.

Solution

Solution in figure 2.7.

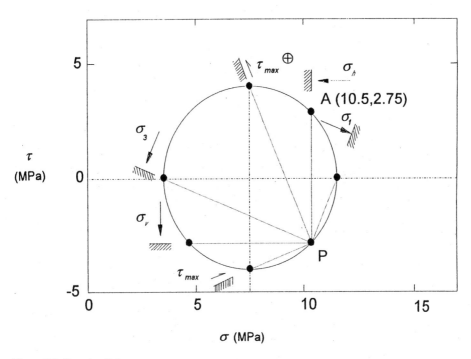

Figure 2.7. Exercise 2.1

The state of deformation

Consider the infinitesimal displacements ∂u, ∂v and ∂w occurring in the direction of the coordinate axes x, y and z, respectively. The following *linear strains* are defined as:

$$\varepsilon_x = \frac{\partial u}{\partial x}$$

$$\varepsilon_y = \frac{\partial v}{\partial y}$$

$$\varepsilon_z = \frac{\partial w}{\partial z}$$

Angular distortions or *shear strains* are defined as:

$$\gamma_{xy} = \frac{\partial v}{\partial x} + \frac{\partial u}{\partial y}$$

$$\gamma_{yz} = \frac{\partial w}{\partial y} + \frac{\partial v}{\partial z}$$

$$\gamma_{xz} = \frac{\partial w}{\partial x} + \frac{\partial u}{\partial z}$$

The strain tensor in the matrix form is, therefore:

$$|\varepsilon| = \begin{vmatrix} \varepsilon_x & \gamma_{xy}/2 & \gamma_{xz}/2 \\ \gamma_{yx}/2 & \varepsilon_y & \gamma_{yz}/2 \\ \gamma_{zx}/2 & \gamma_{zy}/2 & \varepsilon_z \end{vmatrix}$$

Shear strains can be physically interpreted as shown in figure 2.8. After shear deformation, angles θ and β were measured at a point in the material, respectively in the vertical and horizontal direction. These angles can be defined as:

$$\theta = \frac{\partial v}{\partial z} \text{ and } \beta = \frac{\partial w}{\partial y}$$

Following the shear strain definition, the conclusion is:

$$\gamma_{xy} = \theta + \beta$$

Another notation used for strain is ε_{ij}, in which the i and j subscripts take the values of x, y and z. When $i = j$, it refers to linear strain, when $i \neq j$ it refers to shear strain. The shear strains are, therefore:

$$\varepsilon_{ij} = \frac{1}{2}\gamma_{ij}, \, i \neq j$$

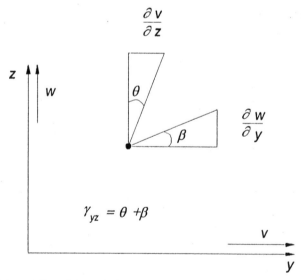

Figure 2.8. Interpretation of shear strains

According to this notation the strain tensor is:

$$|\varepsilon| = \begin{vmatrix} \varepsilon_{xx} & \varepsilon_{xy} & \varepsilon_{xz} \\ \varepsilon_{yx} & \varepsilon_{yy} & \varepsilon_{yz} \\ \varepsilon_{zx} & \varepsilon_{zy} & \varepsilon_{zz} \end{vmatrix}$$

Everything applicable to plane stress is also applicable to plane strain. Therefore, it is possible to demonstrate: principal strains: ε_1, ε_2 and ε_3, strain invariants: J_1, J_2 and J_3 and the Mohr's circle of strain.

Stress-strain relationship

The state of stress, as studied before in this chapter, is represented by a stress tensor having 6 independent stress components. By analogy, the strain tensor has the same number of independent strains.

Assuming linear relationship between stresses and strains, the following matrix equation relates the two sets:

$$\{\varepsilon\} = |C| \ \{\sigma\} \tag{2.14}$$

where: $\{\varepsilon\}$ and $\{\sigma\}$ are vectors whose components are the independent terms of the stress and strain tensors:

$$\{\varepsilon\} = \begin{Bmatrix} \varepsilon_x \\ \varepsilon_y \\ \varepsilon_z \\ \gamma_{xy} \\ \gamma_{xz} \\ \gamma_{yz} \end{Bmatrix} \text{ and } \{\sigma\} = \begin{Bmatrix} \sigma_x \\ \sigma_y \\ \sigma_z \\ \tau_{xy} \\ \tau_{xz} \\ \tau_{yz} \end{Bmatrix}$$

$|C|$ is a 6 x 6 coefficient matrix.

A general solution for equation 2.14 requires the determination of 36 coefficients of the matrix $|C|$. This means 36 different material parameters that would have to be obtained from 36 tests. Since this is not practical, a series of simplifying assumptions on the material behaviour can be formulated to reduce the number material parameters in the $|C|$ matrix. Therefore, as proposed by Hooke back in the 17th century, if one assumes: homogeneity, isotropy and linear-elastic behaviour, the independent coefficients of matrix $|C|$ will be reduced to only 2 constants: Young's modulus E and Poisson's ratio v.

In this particular situation, equation 2.14 can be rewritten as:

$$\{\varepsilon\} = |E| \ \{\sigma\} \tag{2.15}$$

Equation 2.15 is also known as the generalized Hooke's law, in which the coefficient matrix $|E|$ is:

$$|E| = \begin{vmatrix} 1/E & -v/E & -v/E & 0 & 0 & 0 \\ -v/E & 1/E & -v/E & 0 & 0 & 0 \\ -v/E & -v/E & 1/E & 0 & 0 & 0 \\ 0 & 0 & 0 & \dfrac{2(1+v)}{E} & 0 & 0 \\ 0 & 0 & 0 & 0 & \dfrac{2(1+v)}{E} & 0 \\ 0 & 0 & 0 & 0 & 0 & \dfrac{2(1+v)}{E} \end{vmatrix}$$

In the above matrix the inverse of the relation $\dfrac{2(1+v)}{E}$ is called *shear modulus*:

$$G = \frac{E}{2(1+v)} \qquad (2.16)$$

Hooke's law can also be presented in the canonical form, becoming the following system of equations:

$$\varepsilon_x = \frac{\sigma_x}{E} - \frac{v}{E}(\sigma_y + \sigma_z)$$

$$\varepsilon_y = \frac{\sigma_y}{E} - \frac{v}{E}(\sigma_x + \sigma_z)$$

$$\varepsilon_z = \frac{\sigma_z}{E} - \frac{v}{E}(\sigma_x + \sigma_y)$$

$$\gamma_{xy} = \frac{\tau_{xy}}{G} \qquad (2.17)$$

$$\gamma_{yz} = \frac{\tau_{yz}}{G}$$

$$\gamma_{zx} = \frac{\tau_{zx}}{G}$$

Exercise 2.2

An axial compression test was carried out on a cylindrical soil sample. The following stresses were applied: $\sigma_2 = \sigma_3 = 100$ kPa and $\sigma_1 = 300$ kPa. The following results were recorded: $\varepsilon_1 = 6\%$ and $\varepsilon_2 = \varepsilon_3 = -1\%$ (expansion). Obtain the elastic constants E, v and G.

Solution

Introducing values of the applied stresses and the recorded strains into Hooke's law (equation 2.17):

$$0.06 = \frac{300}{E} - \frac{v}{E}(100 + 100)$$

$$-0.01 = \frac{100}{E} - \frac{v}{E}(300 + 100)$$

Solving the above system, one gets $E = 3.8$ MPa and $v = 0.35$. The shear modulus G is obtained from equation 2.16 is $G = 1.4$ MPa.

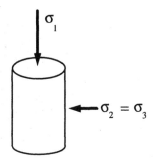

Spherical and deviator tensors

The *spherical tensor* is defined as the matrix $|T^s|$, similar to the stress tensor, but corresponding to a hydrostatic state of stress, in which the shear stresses are nil, and the principal stress equals the mean normal stresses p:

$$|T^s| = |p| = \begin{vmatrix} p & 0 & 0 \\ 0 & p & 0 \\ 0 & 0 & p \end{vmatrix} \tag{2.18}$$

Subtracting the spherical tensor from the stress tensor, one gets the deviator tensor $|T^d|$:

$$|T^d| = |\sigma| - |T^s| = \begin{vmatrix} \dfrac{2\sigma_x - \sigma_y - \sigma_z}{3} & \tau_{xy} & \tau_{xz} \\ \tau_{yx} & \dfrac{2\sigma_y - \sigma_x - \sigma_z}{3} & \tau_{yz} \\ \tau_{zx} & \tau_{zy} & \dfrac{2\sigma_z - \sigma_x - \sigma_y}{3} \end{vmatrix} \tag{2.19}$$

By analogy with stresses, the spherical and deviator strain tensors are defined, respectively as $|D^d|$ and $|D^s|$.

The meaning of these tensors becomes very clear when Hooke's law is rewritten as (suppressing the matrix vertical bars for clarity):

$$T^d = 2G \ D^d \tag{2.20}$$

$$T^s = 3K \ D^s \tag{2.21}$$

where K is called *volumetric, bulk* or *volume change* modulus, also defined as a relationship between the mean stress p and the volumetric strain ε_{vol}:

$$K = \frac{p}{\varepsilon_{vol}} \tag{2.22}$$

Where:

$$\varepsilon_{vol} = \varepsilon_1 + \varepsilon_2 + \varepsilon_3 \tag{2.23}$$

K can also be obtained by:

$$K = \frac{E}{3(1-2v)} \tag{2.24}$$

can be concluded that in an ideal elastic the material, volume change is related to changes in the spherical tensor. Distortions or changes in the material shape are related to the deviator tensor.

Stress-strain behaviour of soils

The application of mathematical models to simulate real soils is an art, because such models do not match entirely with the real behaviour of soils. Departure from theory is due to the fact that soils present hysteresis, non-linearity and irreversible or plastic deformation when loaded. The art resides in choosing the more appropriate model to represent the main aspects of the engineering problem.

Figure 2.9 summarizes the main features of some constitutive models, described as follows.

The Hooke's law, as discussed before, is applicable to homogeneous, linear-elastic materials and those without hysteresis, as shown in figure 2.9a. Any application to real soils must take into account its serious limitations.

In the beginning of the stress-strain curve soils present an approximate elastic behaviour, since strains are recoverable when load ceases. As a consequence, linear elasticity can be assumed if the loading is small enough to keep stresses and strains at a low level. The big advantage of the elastic model is simplicity. Closed-form solutions are available for many situations, as well as charts and graphs, for engineering applications.

If one decides to explore anything beyond elasticity (e.g., figure 2.9b), this will probably lead to the need of a more complex numerical solution on a computer. There are several models that can be used, each one having its advantages and limitations (e.g., Desai and Christian, 1977). A non-linear stress-strain curve could, for instance, be followed in small steps by using small stress increments.

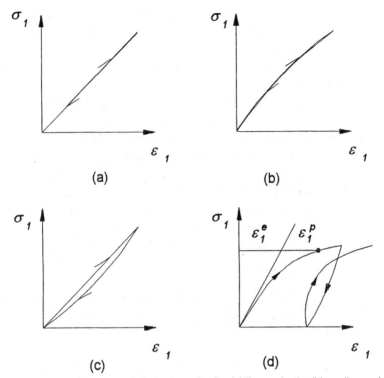

Figure 2.9. Idealization of stress-strain behaviour of soils: (a) linear-elastic; (b) nonlinear-elastic; (c) hysteresis; (d) elasto-plastic

This technique has been used in many practical applications on the design of soil structures.

Since the late sixties, geotechnical engineers became more and more aware of the work carried out at the University of Cambridge on the development of simple, but very powerful, elasto-plastic models. An elasto-plastic assumption is shown in figure 2.9d. The stress-strain curve is represented by an initial linear-elastic behaviour up to a yield stress. The yield stress limits the boundary between elastic and plastic domain. Once this yield limiting stress is reached, plastic strains occur in addition to elastic strains. The difference between plastic and elastic strains is that the former are irrecoverable. Elastic strains, on the other hand, return to zero if the loading ceases. Once an element starts to yield, plastic and elastic are added up to obtain the total strain. This model will be discussed in chapter 14.

Proposed exercises

2.1. What are the equilibrium conditions for the stresses in the vicinity of a point of the continuum?

2.2. What is the strain tensor? Explain the meaning of its components.

2.3. What is the characteristic equation of stresses and what are their roots? Repeat for deformation.

2.4. Explain the physical meaning of the spherical and deviator stress and deformation tensors.

2.5. On a cylindrical soil specimen the following state of stress was applied: $\sigma_1 = 280$ kPa and $\sigma_2 = \sigma_3 = 0$. The resulting strains were $\varepsilon_1 = 6\%$ and $\varepsilon_2 = \varepsilon_3 = -1.5\%$ (expansion). Assuming linear-elastic behaviour, obtain the Young's modulus, Poisson's ratio and volumetric or bulk modulus K.

2.6. A cylindrical sample of saturated soil is tested in axial compression under the same state of stress of the previous exercise. The specimen volume was kept constant throughout the test (i.e., $\varepsilon_{vol} = \varepsilon_1 + \varepsilon_2 + \varepsilon_3 = 0$). The resulting axial strain was $\varepsilon_1 = 5\%$. Evaluate lateral deformations ε_2 or ε_3 and obtain the elastic parameters E, v, G and K.

2.7. Show that in plane strain conditions the intermediate principal stress is given by $\sigma_2 = v(\sigma_1 + \sigma_3)$.

2.8. For the state of stress below obtain the normal and shear stress in a plane inclined $\alpha = 30°$ with horizontal; the principal stresses and their directions; the maximum shear stress and the plane where it acts.

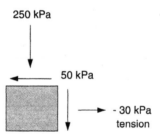

2.9. Repeat previous exercise with the vertical axis turned 30 degrees anti-clockwise.

2.10. The state of strain in an element is given by $\varepsilon_1 = 20\%$ and $\varepsilon_3 = 5\%$. Plot Mohr's circle and work out the maximum angular distortion that can occur in this element. Remember that $\varepsilon_{xy} = \frac{1}{2}\,\gamma_{xy}$.

Chapter 3

Initial stresses in soils

Introduction

This chapter deals with the initial state of stress in soils, which occurs before any external loading has been applied. The effect of water in soils will be studied and the very important concept of effective stress will be introduced.

Initial stresses in soils

Consider the soil profile as shown in figure 3.1. The ground level is horizontal, there are no external loadings and no water in the soil. The soil is considered homogeneous and the apparent total unit weight is γ.

Point A, shown in the figure, is at depth z, where the normal initial vertical stress σ_{v0} is required. Its value can then be calculated considering the weight of a soil column above point A having a unit cross sectional area. Then:

$$\sigma_{v0} = \gamma\ z \tag{3.1}$$

On the other hand, if the soil above A is stratified, i.e., made of n layers, σ_{v0} is given by:

$$\sigma_{v0} = \sum_{i=1}^{n}\gamma_i z_i \tag{3.2}$$

Water in soil

Water enters soils through infiltration and forms aquifers or groundwater. When a soil profile presents several pervious and impervious layers it may lead to the formation of *phreatic*, *confined* or *artesian*, and *perched* aquifers.

Figure 3.2 shows a soil profile where 3 boreholes A, B and C were drilled. Borehole A was carried out through an impervious (hatched) layer until it reaches the pervious bottom layer, in which water is under pressure. The driller observes a sudden rise of water above the ground level in the borehole, when it

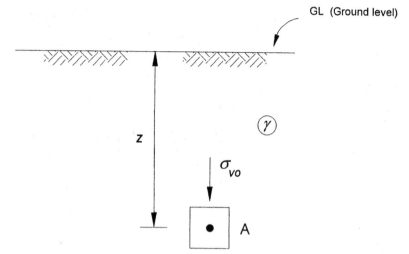

Figure 3.1. Determination of σ_{v0} in dry soil

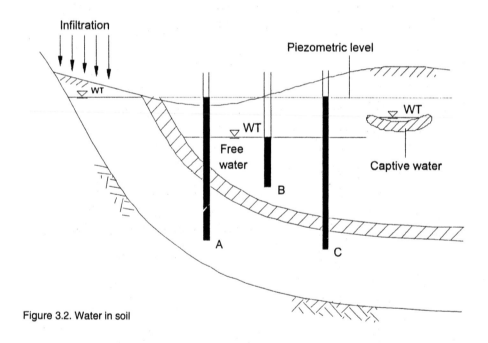

Figure 3.2. Water in soil

reaches this layer. This aquifer is known as *confined* or *artesian*.

Borehole *B* is drilled through a phreatic aquifer, or a free water table. In case *C*, an artesian aquifer is reached. The driller observes water rising in the borehole.

Figure 3.2 also shows a perched water table, occurring over a fine layer of impervious soil. If a borehole is drilled through this perched water table, the water level in the hole may sudden disappear, as soon as it strikes the pervious layer below.

Let's look now to what happens close to the free water table in soils. The well shown in figure 3.3 is excavated to enable a closer look into the soil in the vicinity of the water table. Looking at the vertical walls of the excavation, it can be observed that the soil colour varies as a function of soil moisture. Capillarity enables water to rise above the water table, forming two different zones. The first one up to the point where complete saturation is achieved, is called the *full saturation* capillary zone, and the one above, the *partial saturation* capillary zone. Water pressure within these zones is negative due to capillary tension.

Capillary zones are very important for agronomists, because it is where plants obtain water and other nutrients. For geotechnical engineers, however, the major interest resides below the water table (WT), where initial pore water pressures (u_0), are positive and calculated by the following equation:

$$u_0 = z_w \, \gamma_w \qquad (3.3)$$

where: γ_w is the unit weight of water, taken as 10 kN/m³; z_w depth below WT.

Recently, the interest on the behaviour of residual soils has led to a major concern with what occurs above the WT and the measurements of soil suction in the capillary zones. This may explain mechanisms of slope failures during rainstorms in tropical environments.

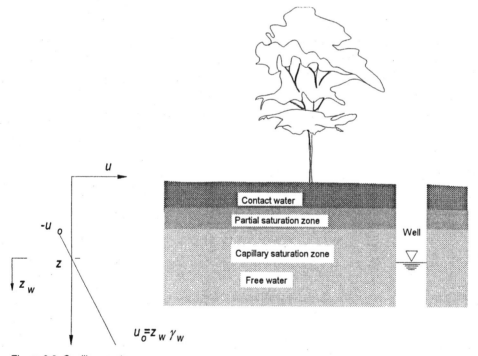

Figure 3.3. Capillary system

Total vertical stresses

The calculation of total vertical stresses in a dry soil has been studied before. If the soil has water, the total vertical stresses are obtained in a very simple way, as shown in figure 3.4. The layers standing above and below the WT are considered separately for the application of equation 3.2.

For point A of figure 3.4:

$$\sigma_{v0} = \gamma \, z_1 + \gamma_{sat} \, z_2$$

Exercise 3.1

For the soil profile in figure 3.5 obtain σ_{v0} at point A considering the WT: (1) at the indicated position; (2) 2 m above ground level.

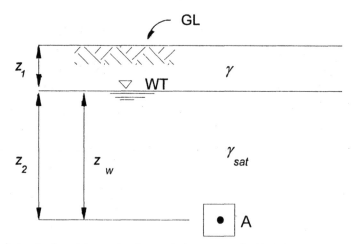

Figure 3.4. Determination of the overburden vertical stress in soil with water

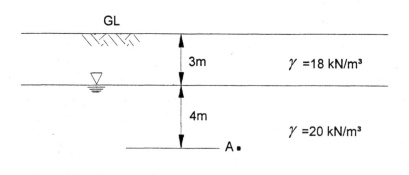

Figure 3.5. Exercise 3.1

Solution

(1) σ_{v0} is calculated considering two soil layers:

$$\sigma_{v0} = \underbrace{3 \text{ m x } 18 \text{ kN/m}^3}_{\text{above WT}} + \underbrace{4 \text{ m x } 20 \text{ kN/m}^3}_{\text{below WT}} = 134 \text{ kPa}$$

(2) If the WT is 2 m above soil level, the water pressure is added, as an additional layer:

$$\sigma_{v0} = \underbrace{2 \text{ m x } 10 \text{ kN/m}^3}_{\text{water}} + \underbrace{3 \text{ m x } 18 \text{ kN/m}^3}_{\text{first layer}} + \underbrace{4 \text{ m x } 20 \text{ kN/m}^3}_{\text{second layer}} = 154 \text{ kPa}$$

The effective stress concept

The effective stress concept was a milestone in civil engineering and a major contribution of Terzaghi, the father of Soil Mechanics.

K. Terzaghi (1883-1963) was a famous engineer and professor, who was born in Austria and later migrated to US. In 1925, he published in Vienna the book *Erdbaumechanik auf den Bodenphychcalische Grundlage*, i.e., Soil Mechanics based on Soil Physics, where he established the effective stress concept based on observation and intuition. According to Terzaghi the average intergranular pressure, which he called *effective stress*, controls the behaviour of saturated soils. Effective stresses are calculated by a very simple equation:

$$\sigma' = \sigma - u \tag{3.4}$$

where: σ' is the effective stress, σ total stress and u, the pore pressure.

The stress tensor in terms of effective stresses becomes:

$$|\sigma'| = |\sigma| - |u| \tag{3.5}$$

where:

$$|\sigma'| = \begin{vmatrix} \sigma'_x & \tau_{xy} & \tau_{xz} \\ \tau_{yz} & \sigma'_y & \tau_{yz} \\ \tau_{zx} & \tau_{zy} & \sigma'_z \end{vmatrix}$$

is the effective stress tensor and the pore pressure tensor is:

$$|u| = \begin{vmatrix} u & 0 & 0 \\ 0 & u & 0 \\ 0 & 0 & u \end{vmatrix}$$

The shear stresses are not altered since water has no shear strength, then:

$$\tau'_{ij} = \tau_{ij}$$

Terzaghi devised a simple experiment to test his principle. He used a water tank

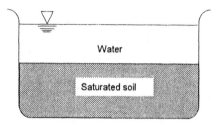

Figure 3.6. Terzaghi's experiment to demonstrate the effective stress principle

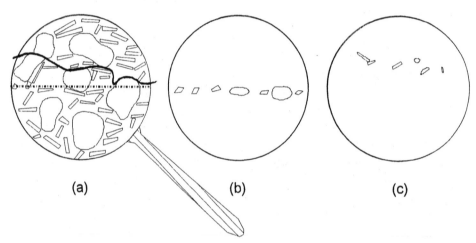

(a)   (b)   (c)

Figure 3.7. Microscopic view of soil particles: (a) A wavy and a horizontal plane sectioning a soil element; (b) Cross section through the horizontal plane; (c) Cross section through the wavy plane

containing saturated loose sand and filled it with water (figure 3.6). The water level was above the top of the soil layer. Then, he raised the water level in the tank and observed that no volume change occurred in the soil. Indeed, if the soil is saturated, the effective stresses do not change at all, as it will be demonstrated in exercise 3.2. As a consequence, no volume change occurred. Terzaghi, then, understood that volume change is a function of effective stresses, and soil behaviour is totally independent of the total stresses.

Effective stresses can be interpreted microscopically, as shown in figure 3.7. A sample of saturated soil has been sectioned by a wavy and a horizontal plane. The cross sections are indicated.

The average total stress σ is given by:

$$\sigma = \sigma'_g a_g + u a_w \tag{3.6}$$

where: σ'_g is the contact stress among soil grains. Its value is very high because of the small contact area; a_w is the ratio of the of the total area corresponding to grain contacts in the wavy section (figure 3.7c), (the value of a_w is very small);

u is the pore pressure; a_w is the ratio of the total area less a_g, then:

$$a_w = 1 - a_g \qquad (3.7)$$

The effective stress σ', acting on the horizontal plane, is approximately equal to the real contact stress between grains multiplied by the contact area:

$$\sigma' \cong \sigma'_g a_g \qquad (3.8)$$

Substituting equations 3.7 and 3.8 in 3.6, it turns out that:

$$\sigma = \sigma' + u(1 - a_g)$$

As a_g is very small, then, $1 - a_g \cong 1$, which results in further simplifications, leading to:

$$\sigma = \sigma' + u$$

Therefore, the effective stresses σ' can be regarded as intergranular stress in saturated soils.

Exercise 3.2

Apply Terzaghi's effective stress principle to check that the vertical effective stress is unchanged in figure 3.6, when the WT is raised.

Solution
Consider a point in the soil layer at depth z below soil level. Make z_w the depth of water in the container and γ and γ_w, respectively, the unit weight of soil and water. Hence:

Total stress: $\qquad \sigma_{v0} = \gamma_w z_w + z\gamma$

Pore pressure: $\quad u_0 = \gamma_w(z_w + z)$

Effective stress: $\sigma'_{v0} = \sigma_{v0} - u_0 = \gamma_w z_w + z\gamma - \gamma_w(z_w + z)$

Simplifying the above expression: $\therefore \sigma'_{v0} = z(\gamma - \gamma_w)$

Since this equation is independent of z_w it means that the effective stress does not change with a change in water level, provided that the WT is above the ground level.

Exercise 3.3

Calculation of vertical total and effective stresses at points A to D on the soil profile of figure 3.8.

Solution

$\qquad$ Point $A \qquad \sigma_{v0} = 2 \times 17 = 34$ kPa

$\qquad\qquad\qquad\qquad u_0 = 0$

$\qquad\qquad\qquad\qquad \sigma'_{v0} = 34$ kPa

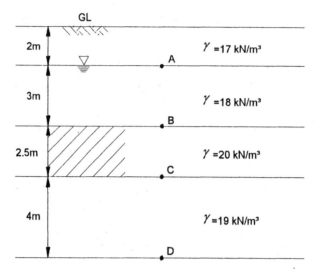

Figure 3.8. Exercise 3.3

Point B $\sigma_{v0} = 2 \times 17 + 3 \times 18 = 88$ kPa

$u_0 = 3 \times 10 = 30$ kPa

$\sigma'_{v0} = 88 - 30 = 58$ kPa

Point C $\sigma_{v0} = 88 + 2.5 \times 20 = 138$ kPa

$u_0 = (3 + 2.5)\, 10 = 55$ kPa

$\sigma'_{v0} = 138 - 55 = 83$ kPa

Point D $\sigma_{v0} = 138 + 4 \times 19 = 214$ kPa

$u_0 = (3 + 2.5 + 4)\, 10 = 95$ kPa

$\sigma'_{v0} = 214 - 95 = 119$ kPa

Another way for working out vertical effective stresses is through the buoyant or the submerged unit weight γ_{sub} or γ' given by:

$$\gamma_{sub} = \gamma_{sat} - \gamma_w \text{ or } \gamma' = \gamma_{sat} - \gamma_w \tag{3.9}$$

This is the easiest way to obtain the vertical effective stresses in most cases. It should not be used when dealing with a perched or artesian aquifer.

Similar calculations of σ'_{v0} in exercise 3.3 can now be done in a single step using γ_{sub} below the WT:

Point A $\sigma'_{v0} = 2 \times 17 = 34$ kPa

Point *B* $\sigma'_{v0} = 34 + 3 (18 - 10) = 58$ kPa

Point *C* $\sigma'_{v0} = 58 + 2.5 (20 - 10) = 83$ kPa

Point *D* $\sigma'_{v0} = 83 + 4 \times (19 - 10) = 119$ kPa

Effective stresses in hydrodynamic conditions

When water flow is occurring in soils, the effective stresses should be obtained through equation 3.4, in which pore pressure *u* is estimated or measured in situ by means of *piezometers*.

The most common type of such instrument is known as the Casagrande piezometer, is shown in figure 3.9. It is installed into the ground by means of a borehole and comprises a porous filter linked to a standing access pipe, in which water is allowed to rise. The bottom part, or the *porous tip* can be made of porous ceramic, porous plastic or a perforated plastic pipe covered with a filter material or simply surrounded by an appropriate geosynthetic. Around the tip a sand filter allows free movement of ground water into or out of the instrument. A plastic access pipe, usually 25 mm in diameter, connects the tip to the ground surface, allowing observation of the water level in the rise pipe. Pore pressure *u* at the tip corresponds to the water column above it.

The next exercise presents an example of the use of the Casagrande piezometer to evaluate artesian pore pressures in the ground and the calculation of vertical effective stresses.

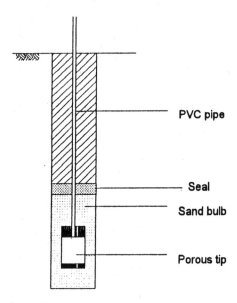

Figure 3.9. Casagrande piezometer

Exercise 3.4

In the soil profile of figure 3.10, piezometers were installed as shown, and an artesian condition was observed in the bottom layer of sand. Obtain σ_{v0}, σ'_{v0} and u_0 at points A, B and C and plot diagrams of these quantities with depth.

Solution

Following the previous exercise 3.3:

$$\text{Point } A \qquad \sigma_{v0} = 2 \times 10 = 20 \text{ kPa}$$

$$u_0 = 2 \times 10 = 20 \text{ kPa}$$

$$\sigma'_{v0} = 0$$

$$\text{Point } B \qquad \sigma_{v0} = 20 + 3 \times 17 = 71 \text{ kPa}$$

$$u_0 = 5 \times 10 = 50 \text{ kPa}$$

$$\sigma'_{v0} = 71 - 50 = 21 \text{ kPa}$$

$$\text{Point } C \qquad \sigma_{v0} = 71 + 2.5 \times 14 + 2 \times 18 = 142 \text{ kPa}$$

$$u_0 = (2 + 2 + 3 + 2.5 + 2) \, 10 = 115 \text{ kPa}$$

$$\sigma'_{v0} = 142 - 115 = 27 \text{ kPa}$$

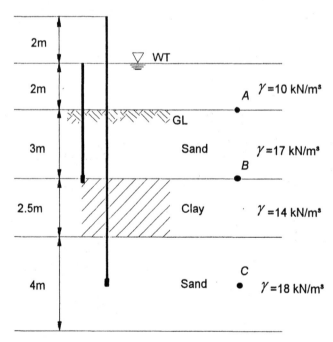

Figure 3.10. Exercise 3.4

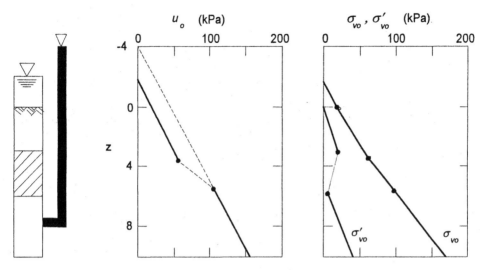

Figure 3.11. Exercise 3.4: plots of $u_0 \times z$ and $\sigma_{v0}, \sigma'_{v0} \times z$

The required plot is shown in figure 3.11. It was obtained from points A, B and C calculated above and some others selected along depth.

Horizontal stresses

The state of stress can only be determined if one knows the stresses occurring on three perpendicular planes. Therefore, in addition to vertical stresses, one must know horizontal stresses (figure 3.12) in order to determine the initial state of stress.

In order to obtain the in situ horizontal effective stress σ'_{h0} it is convenient to define:

$$K_0 = \frac{\sigma'_{h0}}{\sigma'_{v0}} \tag{3.10}$$

K_0 is known as the *coefficient of earth pressures at rest,* defined *only* in terms of effective stresses. This parameter plays a major role in the design of underground structures and excavations, such as in tunneling and underground cavities.

The value of K_0 can be obtained in a laboratory compression test that simulates an *at rest* condition in a soil sample. This implies no significant horizontal deformation during the test.

In order to overcome sampling disturbance effects that may affect the results, a different class of soil tests may be employed: the *in situ tests.* Several different site investigation tools can be used. They vary in complexity and in the amount of disturbance they cause to the ground.

A very simple in situ testing instrument is shown in figure 3.13: the *spade cell*. It consists of an oil filled thin cell made of two welded steel diaphragms. It is inserted into the ground and allowed to rest, while oil pressure is observed. After some time, the gauge pressure will indicate a value for the total horizontal stress into the soil. Experiments with this instrument (e.g., Sully and Campanella, 1989) indicate that it disturbs the stress field into the ground and the measured value has to be corrected to yield σ_{h0}. Pore pressure must be measured independently and, then, the effective horizontal stress σ'_{h0} and the value of K_0 are obtained.

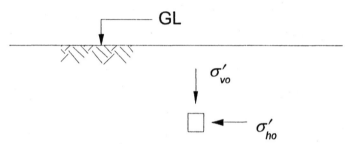

Figure 3.12. Vertical and horizontal effective stresses in a soil element

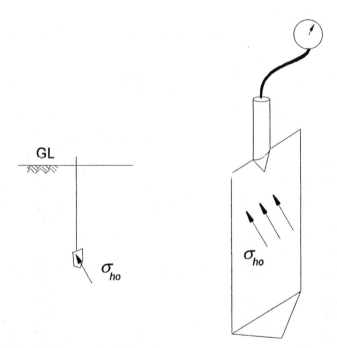

Figure 3.13. Spade cell for measuring horizontal stresses

Exercise 3.5

Obtain the horizontal and vertical effective stresses in points *A*, *B*, *C* and *D* in the soil profile of figure 3.14a and plot them with depth.

Solution

Problem data include unit weights and K_0 which are summarized in table 3.1.

Points *B* and *C* lie on the interface between soil layers, the corresponding values of σ'_{h0} were obtained for two values of K_0, one for the upper and another for the lower soil layer. Figure 3.14b shows the plots. The sudden variation of σ'_{h0} calculated at the transition between two layers of different soils is hypothetical. It cannot occur in real soils, in which transitions from one layer to another are smooth, taking place along a certain distance, say 0.5 to 1 m, at least.

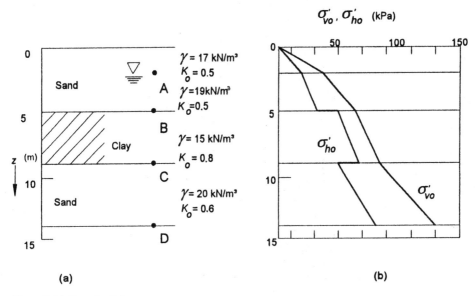

Figure 3.14. Exercise 3.5

Table 3.1. Calculation of initial effective stresses

Point	σ'_{v0} kPa	σ'_{h0} kPa
A	17 x 2 = 34	34 x 0.5 = 17
B	34 + 9 x 3 =61	61 x 0.5 = 30
		61 x 0.8 = 49
C	61 + 4 x 5 = 81	81 x 0.8 = 65
		81 x 0.6 = 49
D	81 + 5 x 10 = 131	131 x 0.6 = 79

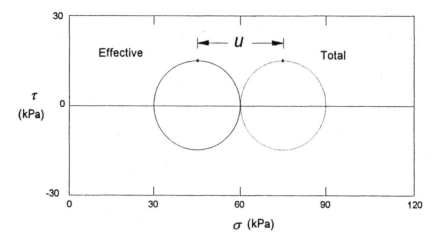

Figure 3.15. Exercise 3.6: Mohr's circles in terms of total and effective stresses

Exercise 3.6

Plot Mohr's circles in terms of total and effective stresses at point B of the previous exercise, in which the effective stresses: are: $\sigma'_{v0} = 61$ kPa and $\sigma'_{h0} = 30$ kPa (the value of σ'_{h0} corresponds to the sand layer in which $K_0 = 0.5$).

Solution
In the soil profile of figure 3.14a the ground surface is horizontal and there are no loadings nearby. Therefore, it can be concluded that σ'_{v0} and σ'_{h0} are principal stresses.

The total principal stresses are obtained adding the pore pressure at point B: $u_0 = 30$ kPa.

$$\sigma_{v0} = 61 + 30 = 91 \text{ kPa}$$

$$\sigma_{h0} = 30 + 30 = 60 \text{ kPa}$$

Mohr's circles are plotted in figure 3.15.

Proposed exercises

3.1. What is a free, an artesian and a perched aquifer?
3.2. Define partial and total capillary zones.
3.3. Why does suction occur in the capillary zone above water table?
3.4. What do you understand by effective stresses? What does effective stress tensor mean?
3.5. Define K_0 and explain how can this coefficient be determined?
3.6. Redo exercise 3.5 with the WT 3 m above ground level. Plot total and effective horizontal and vertical stresses against depth. Plot Mohr's circle for point A.

Chapter 4

Stresses and elastic loading

Introduction

Consider point P of figure 4.1 in which one knows the initial stress tensor $|\sigma_0|$. If loading is applied, it is necessary to determine the final state of stress $|\sigma_f|$ and the increment stress tensor $|\Delta\sigma|$:

$$|\sigma_f| = |\sigma_0| + |\Delta\sigma| \tag{4.1}$$

The study of the effect of loading on the ground was initiated many years ago by the French mathematician Boussinesq applying the theory of elasticity. He worked out closed form solutions for a concentrated load applied on a semi-infinite, linear-elastic, homogeneous, isotropic half-space, having published in Paris in 1885 the book: *Application des potentiels à l'étude de l'equilibre et de mouvement des solides elastiques* (Ed. Gauthiers-Villars).

Once the starting point was set up by Boussinesq, several researchers , managed to solve by integration other problems such as those presented in figure 4.2: line loading, distributed loading on a rectangular surface, triangular loading, embankment loading etc.

The purpose of this chapter is to present a few important solutions frequently employed in geotechnics that can easily be applied through a closed form solution or with the aid of a chart. A comprehensive treatment of this matter has been presented by Poulos and Davis (1974). Other important references are Harr (1966) and Giroud (1975).

Displacements due to loading on the ground can also be calculated through the theory of elasticity. It is not our purpose to cover this subject in detail. The reader is advised to refer to the comprehensive treatise by Poulos and Davis (1974).

The final part of this chapter is dedicated to the introduction of the stress path graphical technique, which will be used throughout the text.

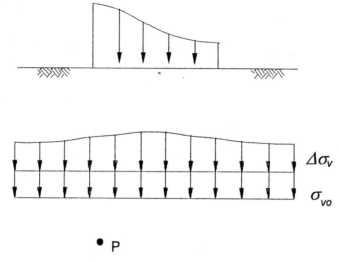

Figure 4.1. Effect of loading on ground level

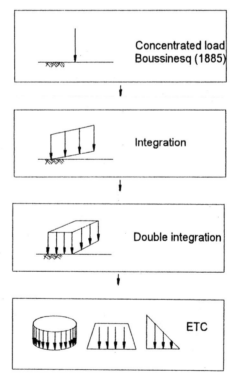

Figure 4.2. Theoretical solutions for stress distribution in soil for various loading patterns, through the integration of Boussinesq's original solution

Boussinesq solution for concentrated loading

Let Q be the concentrated loading (figure 4.3) applied on the surface of a linear-elastic, homogeneous, isotropic half-space. The resulting stress increments at a point with cylindrical coordinates: depth z and radius r are:

$$\Delta\sigma_z = \frac{3Qz^3}{2\pi R^5} \tag{4.2}$$

$$\Delta\sigma_r = \frac{Q}{2\pi}\left[\frac{3r^2 z}{R^5} - \frac{1-2v}{R(R+z)}\right] \tag{4.3}$$

$$\Delta\sigma_\theta = \frac{Q}{2\pi}(1-2v)\left[\frac{z}{R^3} - \frac{1}{R(R+z)}\right] \tag{4.4}$$

$$\Delta\tau_{rz} = \frac{3Qrz^2}{2\pi R^5} \tag{4.5}$$

where: $R^2 = z^2 + r^2$.

An important conclusion regarding the Boussinesq equations is that the stress increments $\Delta\sigma_z$ and $\Delta\tau_{rz}$ are independent of the elastic properties of the material. In other words, independent of the soil type. The radial and hoop stress increments $\Delta\sigma_r$ and $\Delta\sigma_\theta$ are a function of Poisson's ratio, which in most soils lies in the 0.2 to 0.5 range.

These conclusions can be applied to real soils only when elastic behaviour can be assumed, and for a thick soil layer. This restricts the application of the

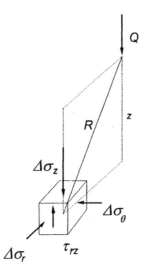

Figure 4.3. Boussinesq's solution for stresses at a point due to a concentrated load

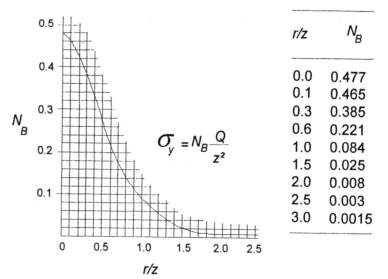

Figure 4.4. Chart for calculation of vertical stresses due to a concentrated load, Boussinesq's solution

theory to the beginning of the stress-strain curve, for small loading, when the safety factor is still very high.

Equation 4.2 can be presented differently, as shown in equation 4.6, allowing the use of the chart in figure 4.4:

$$\Delta\sigma_z = N_B \frac{Q}{z^3} \tag{4.6}$$

where: N_B is the influence factor for $\Delta\sigma_z$ of Boussinesq's solution.

Exercise 4.1

A 1000 kN load is applied on the ground surface. Estimate the final stresses σ_{vf}, σ_{hf} and τ_{vhf} at point P shown in figure 4.5. Take $\nu = 0.5$.

Solution
The coordinates at P are: $z = 3$ m, $r = 3$ m. Therefore: $R = (3^2 + 3^2)^{0.5} = 4.24$ m. Using Boussinesq equations:

$$\Delta\sigma_z = \frac{3 \times 1000 \times 3^3}{2\pi \, 4.24^5} = 9.4 \text{ kPa}$$

$$\Delta\sigma_r = \frac{1000}{2\pi}\left[\frac{3 \times 3^2 \times 3}{4.24^5} - \frac{1 - 2 \times 0.5}{4.24(4.24 + 3)}\right] = 9.4 \text{ kPa}$$

$$\Delta\tau_{rz} = \frac{3 \times 1000 \times 3 \times 3^2}{2\pi \, 4.24^5} = 9.4 \text{ kPa}$$

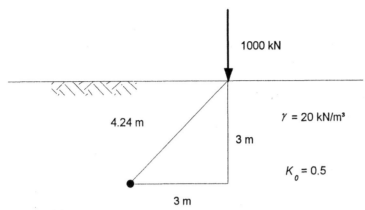

Figure 4.5. Exercise 4.1

The initial stresses are:

$$\sigma_{v0} = 3 \times 20 = 60 \text{ kPa}$$

$$\sigma_{h0} = 0.5 \times 60 = 30 \text{ kPa}$$

The final stresses are:

$$\sigma_{vf} = \sigma_{v0} + \Delta\sigma_z = 60 + 9.4 = 69.4 \text{ kPa}$$

$$\sigma_{hf} = \sigma_{h0} + \Delta\sigma_r = 30 + 9.4 = 39.4 \text{ kPa}$$

$$\tau_{vhf} = \tau_{vh0} + \Delta\tau_{rz} = 0 + 9.4 = 9.4 \text{ kPa}$$

Exercise 4.2

Obtain $\Delta\sigma_z$ for the previous exercise using the chart in figure 4.4:

Solution
The chart is entered with $r/z = 3/3 = 1$. The influence factor is then $N_B = 0.084$. Using equation 4.6:

$$\Delta\sigma_z = 0.084 \frac{1000}{3^2} = 9.3 \text{ kPa}$$

Strip loading

This type of loading (figure 4.6) often occurs in practice, in long retaining wall foundations, or in long shallow foundations of buildings, which apply to the soil the distributed load p per unit area. A shallow foundation can be regarded as sufficiently long when the length L is at least 3 times the breadth B.

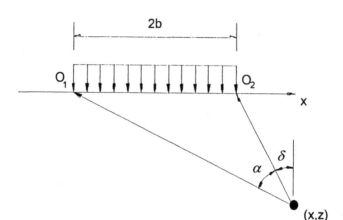

Figure 4.6. Distributed load in a infinite strip

The equations are:

$$\Delta\sigma_z = \frac{p}{\pi}\left[\alpha + \sin\alpha\,\cos(\alpha + 2\delta)\right] \tag{4.7}$$

$$\Delta\sigma_x = \frac{p}{\pi}\left[\alpha - \sin\alpha\,\cos(\alpha + 2\delta)\right] \tag{4.8}$$

$$\Delta\sigma_y = \frac{2p}{\pi}\nu\alpha \tag{4.9}$$

$$\Delta\tau_{xz} = \frac{p}{\pi}\sin\alpha\,\sin(\alpha + 2\delta) \tag{4.10}$$

where: ν is the Poisson ratio of the foundation and α and δ are angles in radians, as defined in figure 4.6.

Circular loading

This situation occurs in tank or chimney foundations having a radius R and applying the unit load p to the soil. Figure 4.7 shows a chart containing isobars, i.e., curves of equal pressure, which allow to obtain $\Delta\sigma_v / p$, the ratio of the vertical stress increment divided by the unit load p, as a function of the rated coordinates x / R and z / R.

Exercise 4.3

Obtain the vertical stress increment at points A and B due to a circular tank foundation (figure 4.8), 6 m in diameter, which applies 240 kPa at the ground level. Point A is under the centre of the tank at a depth of 3 m, B is under the tank periphery, at the same depth.

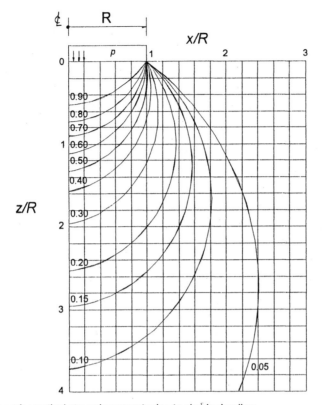

Figure 4.7. Chart for vertical stress increments due to circular loading

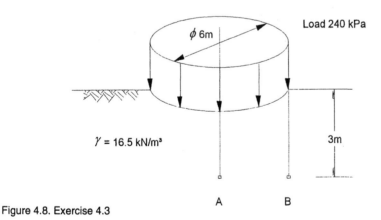

Figure 4.8. Exercise 4.3

Solution

The following ratios are determined first: x/R and z/R, which allow locating the corresponding point in in figure 4.7 chart, and to choose the appropriate iso-bar and the value of $\Delta\sigma_v / p$. Table 4.1 presents computations.

Table 4.1. Calculation of $\Delta\sigma_v$, exercise 4.3

Point	x	z	x/R	z/R	$\Delta\sigma_v/p$	$\Delta\sigma_v$
	(m)	(m)				kPa
A	0	3	0	1	0.64	154
B	3	3	1	1	0.33	79

The pressure bulb

An important concept for practical applications arises from the chart in figure 4.7. The isobar corresponding to 10% of the applied surcharge, in other words, corresponding to $\Delta\sigma_v/p = 0.10$, indicates the soil region that will be significantly affected by the applied loading, and will therefore be called *pressure bulb*, in which most deformation will take place.

The depth reached by the pressure bulb is approximately $2B$, where B is the breadth of the loaded area (figure 4.9a). When the pressure bulb reaches a soft layer below, the foundation soil may present significant deformation. Therefore, an important step in any foundation design is to check the compressibility of the

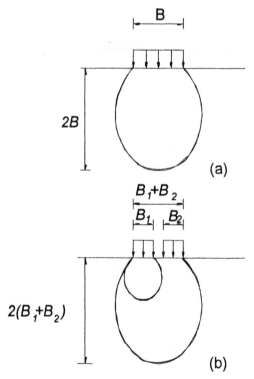

Figure 4.9. (a) Pressure bulb; (b) Interaction between bulbs

soils influenced by the pressure bulb. If the pressure bulb reaches a soft layer, the geotechnical engineer will consider settlements and will decide whether or not this type of foundation is appropriate.

Another important application of the bulb is on the design of a building foundation close to an existing structure. If the bulbs of the new and the existing building interact, as shown in figure 4.9b, the resulting bulb reaches a depth of $2(B_1 + B_2)$, where B_1 is the breadth of the first building, B_2, of the second. Therefore, a check on the layers reached by the bulb is always necessary. If soft soils are present, it should be considered in the design.

Stresses below the edge of a footing

The stress increments below the edge of a rectangular footing with dimensions l and b (figure 4.10), applying the distributed loading p per unit area, are given by the following expressions (Holl, 1940):

$$\Delta\sigma_z = \frac{p}{2\pi}\left[\tan^{-1}\frac{lb}{zR_3} + \frac{lbz}{R_3}(R_1^{-2} + R_2^{-2})\right]$$

$$\Delta\sigma_x = \frac{p}{2\pi}\left[\tan^{-1}\frac{lb}{zR_3} - \frac{lbz}{R_1^2 R_3}\right]$$

$$\Delta\sigma_y = \frac{p}{2\pi}\left[\tan^{-1}\frac{lb}{zR_3} - \frac{lbz}{R_2^2 R_3}\right]$$

$$\Delta\tau_{xz} = \frac{p}{2\pi}\left[\frac{b}{R_2} - \frac{z^2 b}{R_1^2 R_3}\right] \tag{4.11}$$

$$\Delta\tau_{yz} = \frac{p}{2\pi}\left[\frac{l}{R_1} - \frac{z^2 l}{R_2^2 R_3}\right]$$

$$\Delta\tau_{xy} = \frac{p}{2\pi}\left[1 + \frac{z}{R_3} - z(R_1^{-1} - R_2^{-1})\right]$$

where:

$$R_1 = \left(l^2 + z^2\right)^{0.5}$$

$$R_2 = \left(b^2 + z^2\right)^{0.5}$$

$$R_3 = \left(l^2 + b^2 + z^2\right)^{0.5}$$

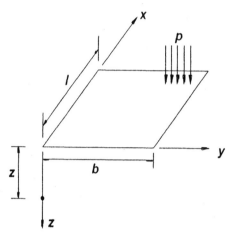

Figure 4.10. Stress increment under the corner of a rectangular loaded area

In the above formulae: $\tan^{-1} x$ is the arc in radians of the angle x; l and b are interchangeable, except in the equations giving $\Delta\tau_{xy}$ and $\Delta\tau_{yz}$.

The vertical stress increment $\Delta\sigma_z$ can also be obtained through the use of charts, as shown in figure 4.11.

Exercise 4.4

Obtain the vertical stress increment $\Delta\sigma_z$ at 5 m depth under the edge of a 6 m x 8 m footing loaded with 300 kPa. Use equations 4.11 and the chart in figure 4.11.

Solution

For p = 300 kPa, z = 5 m, l = 6 m, b = 8 m (alternatively, one may take l = 8 m and b = 6 m with the same results because l and b are interchangeable).

Applying equations 4.11, then:

$$R_1 = \left(6^2 + 5^2\right)^{0.5} = 7.8 \text{ m}$$

$$R_2 = \left(8^2 + 5^2\right)^{0.5} = 9.4 \text{ m}$$

$$R_3 = \left(6^2 + 8^2 + 5^2\right)^{0.5} = 11.2 \text{ m}$$

$$\Delta\sigma_z = \frac{300}{2\pi}\left[\tan^{-1}\frac{6\times8}{5\times11.2} + \frac{6\times8\times5}{11.2}\left(7.8^{-2} + 9.4^{-2}\right)\right] =$$

$$\frac{300}{2\pi}\left[\tan^{-1}0.86 + 0.59\right] = 300 \times 0.21 = 62 \text{ kPa}$$

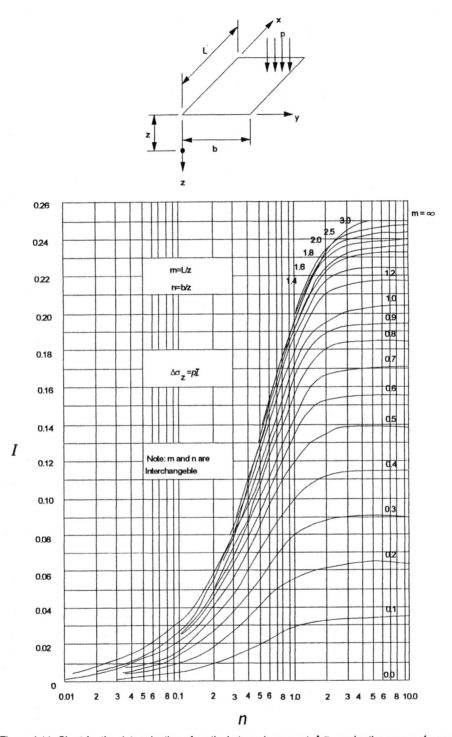

Figure 4.11. Chart for the determination of vertical stress increment $\Delta\sigma_v$ under the corner of a rectangular loaded area (Fadum,1948)

Obtaining $\Delta\sigma_z$ through the chart in figure 4.11, let:

$$m = l/z = 6/5 = 1.2$$

$$n = b/z = 8/5 = 1.6$$

Entering these values into the chart with, one obtains $I = 0.21$. The value of $\Delta\sigma_z$ becomes:

$$\Delta\sigma_z = p\,I \qquad\qquad (4.12)$$

$$\therefore \Delta\sigma_z = 300 \times 0.21 = 62 \text{ kPa}$$

Figure 4.12 presents a chart for a particular footing with $l = 2b$ and the influence factor I for the vertical stress increment can be obtained as a function of the depth ratio. Note that $\Delta\sigma_z$ in the centre of the footing at shallow depth, is bigger than the corresponding value at the corner, but tends to the same value as depth increases.

The charts in figures 4.11 and 4.12 allow the calculation of stresses below the edge of a rectangular footing. It is also possible to use these charts for points situated inside or outside the edge, by applying the principle of superposition of effects, as shown in exercise 4.5. The superposition of effects is only valid in the *elastic* domain. If this is the case, the stresses resulting from two different loadings can be added.

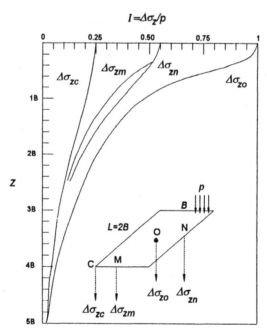

Figure 4.12. Chart for the determination of stresses due to a rectangular loaded area with length twice the width (Giroud, 1975)

Exercise 4.5

Figure 4.13 shows the dimensions of a mat foundation area *HECI* loaded with
$p = 100$ kPa applied at the ground level. Obtain the vertical stress increment at
point *A* at 10 m depth.

Solution
Point *A* is outside the loaded area. Equations 4.11 and the chart in figure 4.11
cannot be used directly. However, assuming elastic domain, the principle of superposition of effects is applied and the problem can be solved. The steps are:
1. The stress increment due to area *ABCD* is determined.
2. The increments due to areas *ABEF* and *AGID* are worked out and subtracted
   from the value obtained in step 1.
3. The stress corresponding to area *AGHF* was deduced twice. It has to be added
   once again.

Table 4.2. Calculation of $\Delta\sigma_z$, exercise 4.5

Area	*l* (m)	*b* (m)	*m*	*n*	*I*	$\Delta\sigma_z$ (kPa)
ABCD	15	20	1.5	2	0.223	22.3
ABEF	5	20	0.5	2	0.135	-13.5
AGID	15	5	1.5	0.5	0.131	-13.1
AGHF	5	5	0.5	0.5	0.085	8.5

$$\Delta\sigma_z = 4.2$$

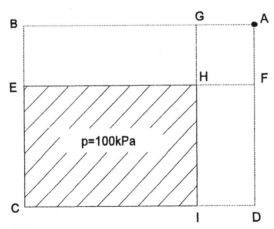

Figure 4.13. Exercise 4.5

Rotation of principal stresses

Comparing the stresses in the soil mass before and after loading, one may observe that the principal stresses have changed their direction, or *rotated*. Mohr's circle is a useful tool to determine the direction of principal stresses, using the techniques described in chapter 2.

Figure 4.14 presents a case of stress rotation where loading has been applied at the ground surface. The crosses represent the magnitude and direction of the principal stresses. The bigger vector corresponds to the major principal stress; the smaller one to the minor. This problem presents a vertical axis of symmetry at the centre of the loaded area. The shear stress increments at this axis are nil, and there is no stress rotation. As the distance to this axis increases, the shear stress also increases, reaching a maximum value below the edge of the loaded area. Note that the major principal stress rotates towards the centre of the loading area. Exercise 4.6 presents an example of how to work out the stress rotation through Mohr's circle.

Exercise 4.6

A certain loading applies at point A in figure 4.15a the following stress increments: $\Delta\sigma_z = 40$ kPa, $\Delta\sigma_z = 30$ kPa and $\Delta\tau_{vh} = 32$ kPa. If the initial stresses are $\sigma_{v0} = 70$ kPa and $\sigma_{h0} = 30$ kPa, obtain the final directions of the principal stresses.

Solution
The final stresses are:

$$\sigma_{vf} = 70 + 40 = 110 \text{ kPa}$$

$$\sigma_{hf} = 30 + 30 = 60 \text{ kPa}$$

$$\tau_{vhf} = 0 + 32 = 32 \text{ kPa}$$

It is necessary to analyse the sign of τ_{vhf} for plotting in Mohr's circle. A sug-

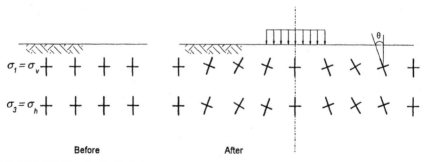

Before   After

Figure 4.14. Rotation of principal stresses

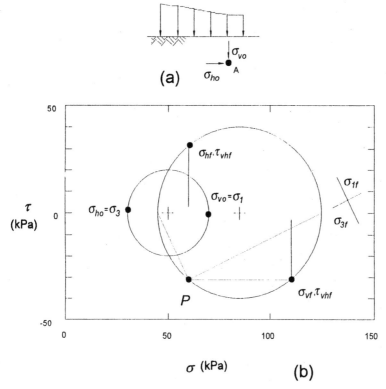

Figure 4.15. Exercise 4.6: Angle of principal stresses rotation through Mohr's circle

gestion is to arbitrate a positive sign, plot Mohr's circle and analyse the final direction of principal stresses. The major principal stress should point towards the centre of the loaded area. If not, the sign of τ_{vhf} has been chosen incorrectly.

In this exercise, the correct sign of τ_{vhf} was found to be negative, as shown in figure 4.15b, and the final principal stress σ_{1f} points to the centre of the surcharge. The steps in figure 4.15b are:

1. Plot the Mohr circle for the following final state of stress: $\sigma_{vf} = 110$ kPa, $\sigma_{hf} = 60$ kPa and $\tau_{vhf} = -32$ kPa.
2. Obtain the pole P, as shown in the figure.
3. Starting at P, draw lines to the points at the circle corresponding to the final principal stresses σ_{1f} and σ_{3f}, which allow the determination of their direction.
4. Analyse if σ_{1f} is pointing to the correct direction. If not, redo from start with opposite sign for τ_{vhf}.

Stress paths

The Mohr circle is an excellent tool for analysing the state of stress in a certain moment of a soil element. However, if the state of stress is changing, as during a laboratory test, the circle is not the appropriate graphical technique.

The need to analyse how stresses are changing becomes very important in some situations as at the modeling of soil behaviour through plasticity models. In fact, if linear-elasticity is assumed, the final state of stress and strain is independent of how they were reached. In plasticity, however, final stresses and strains are a function of how the material has been loaded. Therefore, it is absolutely necessary to know all the previous stresses, or to follow the *stress path*.

A stress path can be obtained by plotting the stress invariants I_1, I_2 and I_3. Each point of this diagram will represent a single state of stress, independent of the direction of the coordinate axes x, y, z. Alternatively, the octahedral stresses could be used in a two-dimensional diagram, because they are related to the invariants. However these solutions are not practical, since the calculation of the invariants or the octahedral stresses is not straightforward.

Two different graphical techniques for stress paths will be used in this book: the MIT plot, originated at the Massachusetts Institute of Technology, USA, (Lambe and Whitman, 1979) and the Cambridge plot, developed at the University of Cambridge, England (e.g., Atkinson and Bransby, 1978).

The MIT plot

The MIT or *s:t* plot has the advantage of matching Mohr's circle and will be preferred in this book. In fact, consider a sequence of states of stresses as shown in figure 4.16a. Take point *A* on the top of the initial circle, point *B* on the following and so forth, until the final point *E*. The MIT stress path is a line connecting the top of each Mohr's circle from points *A* to *E*.

This means taking the following coordinates:

$$s = \frac{\sigma_1 + \sigma_3}{2} \qquad t = \frac{\sigma_1 - \sigma_3}{2} \qquad (4.13)$$

In many cases, as it frequently occurs in laboratory tests, the principal stresses act on the horizontal and vertical planes. Therefore, equations 4.13 can be rewritten as:

$$s = \frac{\sigma_v + \sigma_h}{2} \qquad t = \frac{\sigma_v - \sigma_h}{2} \qquad (4.14)$$

The value of *t* is taken as positive when the vertical stress is equal to the major principal stress; if this not the case it is taken as negative.

Plotting a point in the *s:t* diagram can be done in two ways:

One method is to calculate *s* and *t* coordinates through equations 4.13 or 4.14 and plot a (*s*, *t*) Cartesian graph from data points. This method is preferred when a computer is used.

The second or the *Locus method*: refer to plotting the *loci* of points having

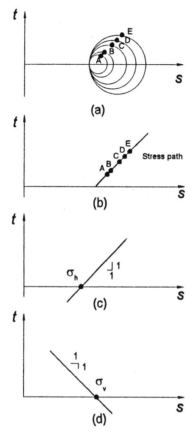

Figure 4.16. Stress path in the MIT diagram; (a) Mohr's diagram; (b) *s:t* diagram; (c) Loci for points with $s = \sigma_h$; (d) Loci for points with $s = \sigma_v$

the same σ_v and σ_h. This is a very useful technique for manual plotting and interpretation of stress paths. It should be practiced by the reader before going beyond this chapter. The loci of all points in the *s:t* plot having a certain σ_v and σ_h are shown in figures 4.16c and 4.16d.

The following simple rules should be memorized.
– The locus of points having the same σ_h is a line that crosses the abscissae axis at $s = \sigma_v$ sloping 1:1 to the right (figure 4.16c).
– The locus of points having the same σ_v is a line that crosses the abscissae at $s = \sigma_h$ sloping 1:1 to the left (figure 4.16d).
– The locus of points having $\sigma_v = \sigma_h$ is a line coinciding with the horizontal *s* axis (figure 4.17a). This axis is also known as the *hydrostatic* axis, since all shear stresses are nil.
– The locus of points with a constant *t / s* ratio, or constant value of $K = \sigma_h / \sigma_v$ is an inclined line shown in figure 4.17b.

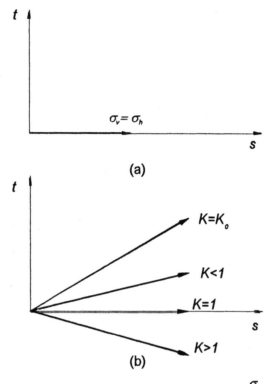

Figure 4.17. (a) Loci for points with $\sigma_v = \sigma_v$; (b) Loci for points with $K = \dfrac{\sigma_h}{\sigma_v} = $ constant

Exercise 4.7
Plot the following points on the $s{:}t$ diagram utilizing the locus method.
(a) $\sigma_v = 200$ kPa, $\sigma_h = 100$ kPa;
(b) $\sigma_v = 150$ kPa, $\sigma_h = 100$ kPa;
(c) $\sigma_v = \sigma_h = 100$ kPa

Solution
(figures 4.18a to 4.18c):
(a) Using only the s axis, locate the locus of points having s = σ_v, which is a line inclined 1:1 to the left. Then, repeat for s = σ_h. The intersection of the two loci is the required point.
(b) As above.
(c) The loci cross on the s axis, as shown in the figure.

Exercise 4.8
Plot the stress paths in a $s{:}t$ diagram for the following loading cases:
(a) Initial stresses $\sigma_v = \sigma_h = 200$ kPa, σ_h remains constant while σ_v increases up to 600 kPa;

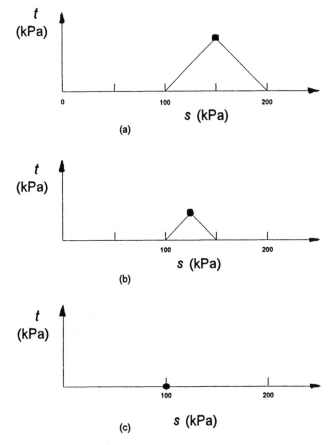

Figure 4.18. (a) Exercise 4.7a; (b) Exercise 4.7b; (c) Exercise 4.7c

(b) Same initial stresses, σ_v remains constant while σ_h increases up to 600 kPa;
(c) Same initial stresses, σ_v remains constant while σ_h decreases to 100 kPa;
(d) Same initial stresses, σ_v and σ_h increase in a constant ratio of $\Delta\sigma_h / \Delta\sigma_v = 3$.

Solution
Cases (a), (b) and (c) are straightforward and are plotted in figure 4.19. Case (d): starts from $s = \sigma_v = \sigma_h = 200$ kPa. A second point at the stress path is obtained as follows: choose arbitrarily an increment $\Delta\sigma_h$, say $\Delta\sigma_h = 300$ kPa. Then, obtain the correspondent $\Delta\sigma_v$ from the given condition: $\Delta\sigma_h = 3 \times \Delta\sigma_v$, $\therefore \Delta\sigma_v = \Delta\sigma_h / 3 = 100$ kPa. Plot a point having $\sigma_v = 200 + 100 = 300$ kPa and $\sigma_h = 200 + 300 = 500$ kPa. This point belongs to the required stress path.

Total and effective stress paths

According to the definition of total and effective stresses, total stress path (*TSP*)

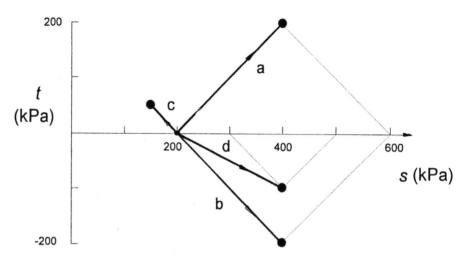

Figure 4.19. Exercise 4.8

and effective stress path (*ESP*) are defined in the *s:t* or *s':t'* diagram. The effective stress plots have the following coordinates:

$$s' = s - u \qquad\qquad t' = t \qquad\qquad (4.15)$$

Exercise 4.9

In a MIT diagram, plot point *B* from exercise 3.6, having the following stresses $\sigma_{v0} = 91$ kPa, $\sigma_{h0} = 60$ kPa and $u_0 = 30$ kPa.

Solution

The total stress plot has been discussed before. The corresponding point in effective stresses *B'* is plotted in figure 4.20 on the same diagram by translating the point to the left by a quantity equal to the pore pressure *u*.

The Cambridge type diagram

The Cambridge type plot differs from MIT's because it includes the intermediate principal stress σ_2 (e.g., Atkinson and Bransby, 1978). The coordinates are *p* and *q* (figure 4.21), which are related to the stress invariants (equations 2.11 and 2.13), and with octahedral stresses σ_{oct} and τ_{oct}, through the following equations:

$$p = \sigma_{oct} \qquad\qquad q = \frac{3}{\sqrt{2}}\tau_{oct} \qquad\qquad (4.16)$$

A simplification in the definition of *q* is used in 2 dimensional applications:

$$q = \sigma_1 - \sigma_3 \qquad\qquad (4.17)$$

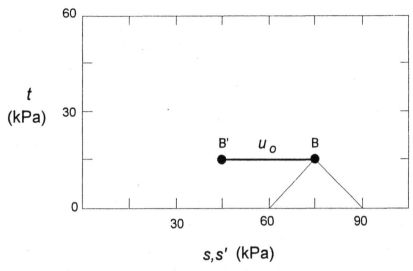

Figure 4.20. Exercise 4.9

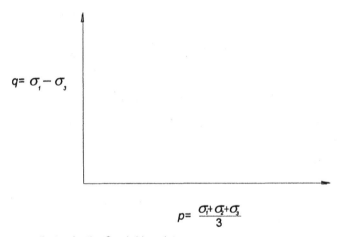

Figure 4.21. *p:q* coordinates for the Cambridge plot

In effective stresses, analogous definitions apply to the Cambridge plot:

$$p' = p - u \qquad q' = q \tag{4.18}$$

In plane strain, σ_2 depends on the remaining principal stresses. If the soil is assumed to be elastic, it can be verified that $\sigma_2 = v(\sigma_1 + \sigma_3)$. On the other hand, in cases where the soil volume is kept constant during the deformation process, as in an undrained test on a saturated sample, Poisson's ratio is $v = 0.5$. In this case, it turns out that $p = s$. In case of drained behaviour, if the soil Poisson's ratio is close to 0.2, then $p' = 0.4(\sigma_1' + \sigma_3')$, and therefore s' is close to p'. For these reasons the $s':t$ plot is preferred for plane strains situations.

Proposed exercises

4.1. What are the soil conditions in which Boussinesq's equations could be applied?

4.2. Define pressure bulb and explain its practical importance.

4.3. A retaining wall is constructed on a sandy soil and transmits a pressure of 500 kPa through a 4 m wide strip footing. Soil data are: $\gamma = 20 \text{ kN/m}^3$, $K_0 = 0.6$ and the WT is 1 m below ground surface. Plot the *TSP* and *ESP* both in the MIT and Cambridge diagrams for a point located 4 m below the ground.

4.4. Compare the stress distribution with depth of: (a) a concentrated load of 4000 kN and (b) the same load distributed through a 4 m x 4 m footing. Plot the results.

4.5. What is the principle of superposition of effects and what conditions for it considered valid?

4.6. The centre of a rectangular area on the ground surface has the following co-ordinates in metres: (0,0), and the borders (6,15). A distributed loading of 400 kPa is applied. Work out the vertical stress increments at a 15 m depth in the points having the following coordinates: (0,0), (0,15), (6,0) and (10,25).

4.7. Point *P* is in a dry sand layer of soil without any loading at the ground level. Then, a concentrated 1000 kN load is applied as shown in the figure below. A second 1500 kN is applied as well. The forces and point *P* are in the same plane. The sand layer presents the following properties: $\gamma = 20 \text{ kN/m}^3$, $K_0 = 0.5$, $v = 0.3$ Obtain: (a) the initial stresses; (b) the stress increments through Boussinesq equations; (c) the rotation of principal stresses; (d) the stress path in point *P* plotted in a *p:q* diagram.

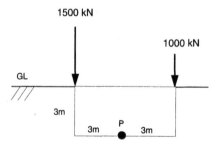

4.8. Plot a MIT diagram the *ESP* for a material loaded with $K_0 = 0.60$. Repeat for $K_0 = 1.1$.

4.9. Plot a *s:t* diagram the following *ESP*'s on a sample of dry sand: (a) initial stresses $\sigma_v = 100$ kPa and $K_0 = 0.6$; (b) σ_h is kept constant while σ_v increases up to 250 kPa; (c) σ_h is kept constant and a decrement $\Delta\sigma_v$ of -30 kPa is applied; (d) σ_h is kept constant and a decrement $\Delta\sigma_v$ of -30 kPa is applied.

Chapter 5

Soil hydraulics

Introduction

This chapter deals with water flow in soils and its importance to engineering structures. In earth dams, for instance, the geotechnical engineer wishes to know the amount of water that will be lost due to seepage through the embankment dam and its foundation; in an industrial waste disposal pond, one wishes to create a nearly impervious soil barrier to avoid ground water contamination. Therefore, the engineer must select the appropriate soil type that has the required permeability.

Flow regime in soils

The theoretical bases of the flow of liquids in conduits were established by Reynolds in 1883 (Reynolds O., 1883, *An experimental investigation of the circumstances which determine whether the motion of water shall be direct or sinuous and of the law of resistance in parallel channels*, Phil Trans, The Royal Society, London). The well-known Reynolds experiments, a subject dealt with in all Fluid Mechanics text-books, showed that the flow regime in pipes can be laminar or, under certain conditions, turbulent. His experiment is sketched in figure 5.1a and consisted of permitting water to flow in a transparent pipe and using a dye to evaluate the flow regime. If the dye trace was a straight line, the flow would be laminar, if not it would be turbulent. Reynolds varied the pipe diameter D and the pipe length L, the head h between two reservoirs and observed the flow velocity v. Typical results are shown in figure 5.1b, where the hydraulic gradient $i = h / l$ is plotted against the velocity of flow v.

Reynolds has also shown (figure 5.1b) that there is a critical velocity of flow v_c below which the flow regime is laminar and there is proportionality between the hydraulic gradient and the flow velocity. On the other hand, for flow rates greater than v_c, there is no such proportionality and the flow is turbulent. Rey-

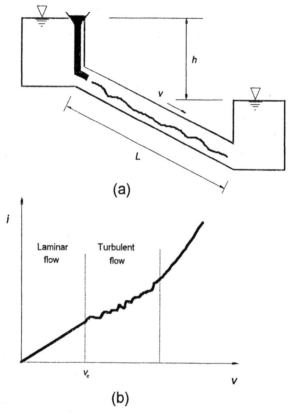

Figure 5.1. Reynolds experiments: (a) Equipment; (b) Results

nolds also worked out an equation that relates v to other quantities:

$$\Re = \frac{v_c D \gamma}{\mu \gamma} \tag{5.1}$$

where: $\Re$ is the Reynolds number, a pure number, equals to 2000; v_c is the critical velocity of flow; D is the pipe diameter; γ is the unit weight of the fluid; μ is the viscosity of the fluid; g is the acceleration of gravity.

Entering values correspondent to water at 20°C into equation 5.1, one gets v_c in m/s as a function of the pipe diameter D in metres:

$$v_c = \frac{28 \times 10^{-4}}{D} \tag{5.2}$$

Pore diameter in soils can be regarded as less than 5 mm. Entering this value into equation 5.2, one obtains $v_c = 0.56$ m/s, which is well beyond the range of velocities that occur in soils. In fact, seepage in soils occurs in velocities well below the critical one and it can be concluded that the flow regime is laminar and there is proportionality between flow velocity and hydraulic gradient. The coefficient of proportionality k is named *permeability* or *hydraulic conductibility*.

$$v = k \, i \tag{5.3}$$

Darcy's law

Equation 5.3, which was derived from Reynolds theory, was indeed obtained experimentally nearly 30 years before in 1856 by the French engineer H. d'Arcy who published the book *Les fontaines publiques de la ville de Dijon* (Ed Dalmon, Paris), and came to be known as *Darcy's law*. Only for didactic purposes, however, this subject has not been introduced chronologically here.

Darcy's experiments are sketched in figure 5.2 and consisted of allowing seepage of water through a soil sample of area A and length L by means of two reservoirs, whose difference in level was h, and measuring the flow rate or discharge Q. The results indicated that flow velocity $v = Q/A$ is proportional to the hydraulic gradient $i = h/L$, according to equation 5.3.

Measuring soil permeability

The permeability of soils can be measured through laboratory and in situ tests. This chapter only covers the most common types: the laboratory tests with constant or variable head permeameters. The former is shown in figure 5.3a and was the type employed by Darcy. It includes two reservoirs where the water level is kept constant and the following measurements are obtained: head h, soil sample length L and sample cross sectional area A, flow rate Q. The value of permeability is given by:

$$k = \frac{QL}{Ah} \tag{5.4}$$

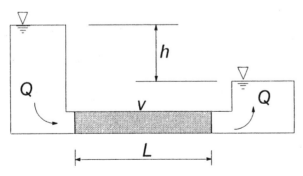

Figure 5.2. Darcy's experiments

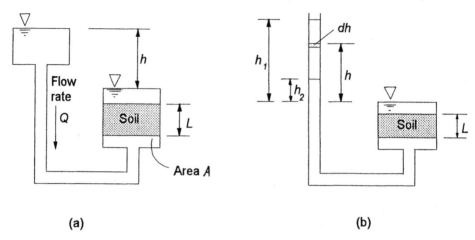

(a)   (b)

Figure 5.3. (a) Constant head permeameter; (b) Variable head permeameter

Exercise 5.1

A test carried out in a constant head permeameter yielded the following data: (a) volume of water that flowed during 500 seconds 0.034 m³; (b) $h = 2$ m, $L = 0.20$ m, $A = 0.04$ m². Obtain the permeability k.

Solution

The discharge was $Q = 0.034/500 = 6.8$ x 10^{-5} m³ /s, using equation 5.4:

$$k = \frac{6.8 \times 10^{-5} \times 0.20}{0.04 \times 2} = 1.7 \times 10^{-4} \text{ m/s}$$

In the variable head permeameter, the soil sample is tested as shown in figure 5.3b. The water level varies constantly during the test in a glass pipe or burette of internal cross section area a. If h_1 and h_2 are measurements of head in times t_1 and t_2 the permeability is given by:

$$k = \frac{aL}{A(t_2 - t_1)} \ln \frac{h_1}{h_2} \tag{5.5}$$

Exercise 5.2

Deduce equation 5.5 for a variable head permeameter.

Let dV be the elementary volume of height dh, in which the volume is $dV = a \, dh$.

Applying Darcy's law $v = k \, i$ and considering that flow velocity v is equal to $v = V/(A \, dt)$, it turns out that $dV = k \, i \, A \, dt$.

Solution

It comes, then:

$$a \, dh = k \, i \, a \, dt$$

$$a\, dh = k\, (h/L)\, A\, dt$$

$$\therefore \frac{dh}{h} = \frac{kA}{aL} dt$$

Integrating between h_1 and h_2 and t_1 and t_2:

$$\int_{h_1}^{h_2} \frac{dh}{h} = k\frac{A}{aL} \int_{t_2}^{t_1} dt$$

$$\therefore k = \frac{aL}{A(t_2 - t_1)} \ln\frac{h_1}{h_2}$$

Values of permeability

Typical permeability values for soils are presented in table 5.1. Pervious or *free draining soils* have permeabilities greater than 10^{-7} m/s. Otherwise, it is a soil presenting impervious or impeded drainage.

Figure 5.4 presents permeability correlations for several types of soils (Vargas, 1977) employing a logarithmic equation of the type $\log k = f(e)$, where e is the void ratio. As this figure includes a wide range of different soils from various origins, it can be concluded that correlations of this type are generally applicable to all soils. Another correlation for the undisturbed and remoulded soils is shown in figure 5.5.

A useful application of such correlations is in the evaluation permeability change with depth. The steps are: first, from permeability test data obtain the correlation $\log k = f(e)$. From void ratio e versus depth z data, work out the desired relationship $\log k = f(z)$.

In sands, an indirect manner of estimating their permeability is Hazen's equation (Hazen, A., 1911, Discussion on *Dams on sand foundations*, Transactions ASCE, vol. 73) applicable to clean uniform sands without fines:

$$k = C\, D_{10}^2 \tag{5.6}$$

where: k is the permeability in m/s, D_{10} is the *effective diameter* in metres, ob-

Table 5.1. Permeability of soils

Soil	Permeability	Soil type	k (m/s)
Free draining soils	high	gravels	$> 10^{-3}$
	high	sands	10^{-3} to 10^{-5}
	low	silts and clays	10^{-5} to 10^{-7}
Impervious soils	very low	clays	10^{-7} to 10^{-9}
	practically impervious	clays	$< 10^{-9}$

Permeability (m/s)

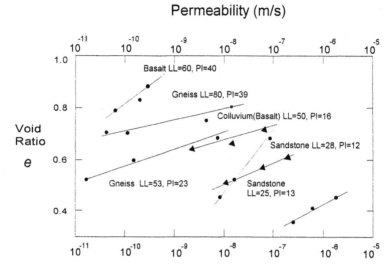

Figure 5.4. Results of permeability tests in residual soils (Vargas,1977)

Permeability (m/s)

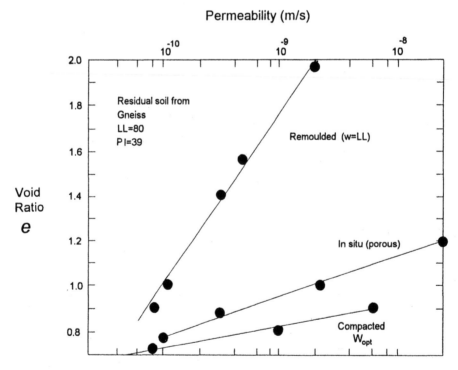

Figure 5.5. Data on permeability versus log void ratio for the same soil in different states (Vargas,1977)

tained from the grain size distribution curve, corresponding to such a diameter that only 10% of the material are retained. The empirical coefficient C is taken as 0.01.

Potentials

Water movement, as well as heat or electricity, is controlled by its state of energy, or its potential Ψ. Different forms of energy can be studied separately, like the kinetic and the potential energy, studied in Physics. Water movement can be studied as a result of a difference in potential, since equilibrium is reached when potential is minimum.

Potentials are related to a reference value. $\Psi_0 = 0$ is assigned for water in normal conditions of temperature (0 °C) and pressure (atmospheric pressure at sea level).

Units commonly adopted for potential are energy per unit of mass, unit of volume or unit of weight. The following recommendations apply when choosing how to express the potential:

– *Energy per unit of mass:* The *SI* unit for energy is the joule (J) corresponding to the work of a 1 N force travelling 1 metre. In engineering it is convenient to utilize the kJ, which, rated by the unit of mass, yields kJ/kg.

– *Energy per unit of volume:* The corresponding *SI* unit is kJ/m, but since kJ = kNm, then: kJ/m = kNm/m = kPa. Units of pressure are obtained.

– *Energy per unit of weight*: It can be seen that it has length dimensions, or the height of a liquid column. When expressed this way, this form of energy is the *hydraulic head (h)*.

The total potential of water Ψ_t can be studied (Reichardt, 1985) as the sum of several components: the kinetic Ψ_c, the piezometric Ψ_p, the altimetric Ψ_a, the thermal Ψ_k, and the matric Ψ_m:

$$\Psi_t = \Psi_c + \Psi_p + \Psi_a + \Psi_k + \Psi_m \tag{5.7}$$

The kinetic component Ψ_c is proportional to the square of the flow velocity v_c. Values of flow velocity v that occur in soils are very small, therefore the kinetic potential becomes negligible and can be disregarded in soil mechanics applications.

The piezometric component Ψ_p is the difference between the pore water pressure in a soil element and the reference potential Ψ_0, i.e., the atmospheric pressure. It is equal to the pore pressure u in the soil element.

The altimetric component Ψ_a, also called gravitational component, is the potential energy due to gravity, equal to mgz, where m is the mass, g the acceleration of gravity, and z the elevation.

The thermal component Ψ_k can also be neglected due to the very small temperature gradients occurring in soils, therefore water flow can be assumed to be an isothermal process.

The matric component Ψ_m is the result of capillary and adsorption effects in soils due to interaction of water and solid soil particles. The interaction forces attract and fix water in soil, decreasing its potential relatively to free water, due to capillary action. This potential can be significant for elevations above the ground water surface in the capillary zone, but is nil below the water table. Any theoretical treatment of this potential is very difficult and experimentation is the only practical way of assessing it (Reichardt, 1985). The matric potential has considerable importance in the study of flow is unsaturated soils and recent studies of slope stability in residual and unsaturated soils include its effect.

Hydraulic head

The hydraulic head (h) is the energy of water per unit of weight. It has units of length. Neglecting the kinetic, thermal and the matrix potential, equation 5.7 can be expressed in terms of hydraulic heads:

$$h_t = h_p + h_a \qquad (5.8)$$

where: h_t is the *total* head, h_p the *piezometric* head and h_a the *altimetric* head.

The piezometric head is given by:

$$h_p = u / \gamma_w \qquad (5.9)$$

where: u is the pore pressure and γ_w the unit weight of water.

The altimetric head is equal to the elevation of the soil element relatively to a reference. Exercises 5.3 to 5.6 give examples of calculation of heads.

Exercise 5.3

Obtain the head diagram (elevation x head) for points 1 and 2 of the water tank shown in figure 5.6.

Solution

Values of the piezometric, altimetric and total head are as follows:

Points	Heads		
	Altimetric	Piezometric	Total
1	h_{a1}	h_{p1}	$h_{a1} + h_{p1} = h_t$
2	h_{a2}	h_{p2}	$h_{a2} + h_{p2} = h_t$

The total head in points 1 and 2 is equal to h_t. Plotting the diagram of heads (figure 5.6a), there is no change in the total head for any point in the tank, which is a condition for no flow.

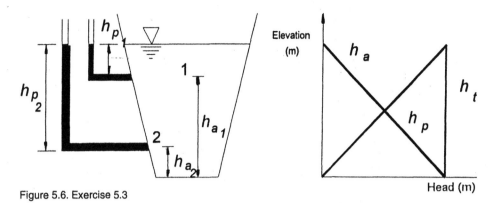

Figure 5.6. Exercise 5.3

Capillary tubing

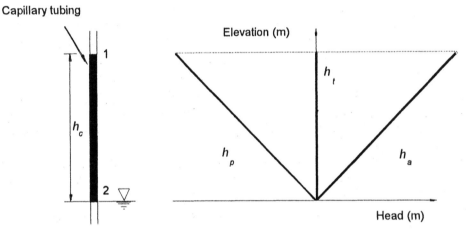

Figure 5.7. Exercise 5.4

Exercise 5.4

Plot the head diagram for the capillary tube of figure 5.7.

Solution

Heads for points 1 and 2 are presented in the following table:

Points	Heads		
	Altimetric	**Piezometric**	**Total**
1	h_c	$-h_c$	$h_c + (-h_c) = 0$
2	0	0	$0 + 0 = 0$

Exercise 5.5

Obtain the head diagram for the soil sample of figure 5.8.

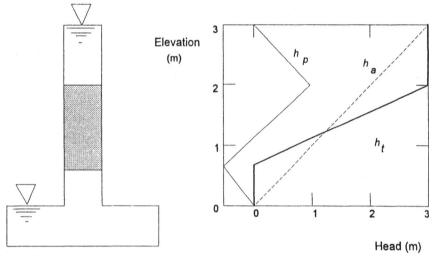

Figure 5.8. Exercise 5.5

Solution

The following steps were adopted for plotting the head diagram:

(a) Plot the altimetric head: the scales for the abscissae and ordinates are the same, the altimetric head will be a line inclined 45°, as indicated in the figure.

(b) Plot the piezometric head: use equation 5.7. Start at the water surface where the pressure u is nil.

(c) Finally, the head diagram is obtained by adding up the previous heads.

Note that the total head only varies within the soil sample where all energy losses take place.

Exercise 5.6

Plot the head diagram for the situation shown in figure 5.9.

Solution

The solution for this case is analogous to the previous exercise.

Exercise 5.7

In exercise 5.5, obtain flow velocity v, assuming that the soil sample has a permeability of 3×10^{-5} m/s.

Solution

Applying Darcy's law, the value of hydraulic gradient is given by $i = h/L$, where h is the total head applied to the sample, equal to 3 m (figure 5.8), L is the sample length, equal to 1.5 m. Thus $i = 3/1.5 = 2$. The flow velocity is given by:

$$v = ki = 3 \times 10^{-5} \times 2 = 6 \times 10^{-5} \text{ m/s}$$

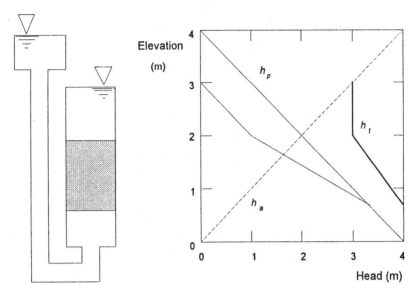

Figure 5.9. Exercise 5.6

Seepage forces

When water flows in soils it loses energy from friction with soil particles. This, in turn, applies a reaction force to soil elements. This is the origin of *seepage force F_p*. Plotting the head diagram of flow in a soil sample, one can work out the amount of energy being dissipated, and therefore, obtain the seepage force acting on the sample.

In order to assess the value of F_p, consider the soil sample of figure 5.10a. The forces acting on the sample, known as boundary forces, can be divided into *submersion* and *seepage* forces. The latter is due to the dissipation of hydraulic head h and can be obtained through $F_p = h\,\gamma\,A$, according to figure 5.10b. Consider, now, the value of the seepage force per unit volume $F_p\,/\,V$:

$$\frac{F_p}{V} = \frac{h\,\gamma_w\,A}{L\,A} = \frac{h}{L}\gamma_w = i\,\gamma_w$$

Hence:

$$\frac{F_p}{V} = i\,\gamma_w \tag{5.10}$$

Therefore, the seepage force per unit volume is proportional to the hydraulic gradient i.

Seepage forces play an important role in geotechnics. In slopes, for instance, they have an enormous influence on the stability and can be taken into account in either of these ways:

1. Using total unit weights for soils and boundary forces;

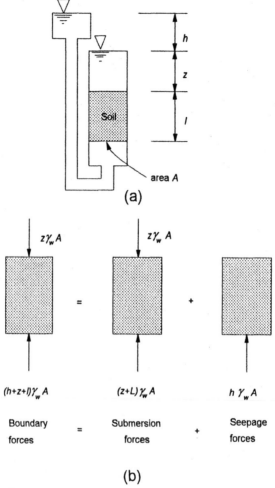

Figure 5.10. Seepage forces

2. Using submerged unit weights for soils and seepage forces.

In slope stability analysis, method (1) is generally recommended (e.g., Lambe and Whitman, 1979). Both methods give the same answer. In fact, figure 5.11a shows the forces acting on a soil element during seepage. The resultant force F is:

$$F = \gamma_t LA + z\gamma_w A - (h + z + L)\gamma_w A$$

$$F = (\gamma_w + \gamma')LA - h\gamma_w A$$

$$F = \gamma' LA - h\gamma_w A \tag{5.11}$$

On the other hand, obtaining the same force F from figure 5.11b:

$$F = \gamma' LA - h\gamma_w A \tag{5.12}$$

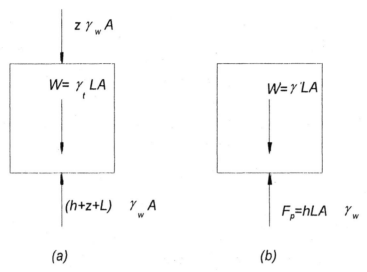

Figure 5.11. Seepage forces

Since equations 5.11 and 5.12 are exactly the same, it can be concluded that the forces on a soil element can be obtained utilizing either of these methods.

Liquefaction

Liquefaction may occur in fine granular soils when the pore pressures rise up to a very high value and reduce the effective stresses to a negligible value. The intergranular pressures turn to zero and there will be no shear strength. The soil behaves as a liquid. This condition is also referred to as the *quick* condition.

Liquefaction has been observed in fine and loose granular soils during earthquakes (e.g., Prakash, 1981) and it is a major concern in foundation design in earthquake zones, as in western South and North America and in Japan. Cyclic strains lead to pore pressure build up which, in turn, cause this phenomenon.

Liquefaction can also occur due to seepage in the ascending direction, as shown in figure 5.12, when seepage force F_p reaches the value of the submerged weight W of the soil element. Thus, taking $W = F_p$ as the critical situation, it is possible to define the *critical hydraulic* gradient i_c, given by:

$$i_c = \frac{\gamma'}{\gamma_w} \tag{5.13}$$

For the majority of soils its submerged unit weight is not much greater than the unit weight of water. Therefore, assuming the ratio $\gamma'/\gamma_w \cong 1$, the critical hydraulic gradient i_c is approximately equal to 1. This situation has to be avoided at any cost in practice, as shall be discussed later in exercise 5.11.

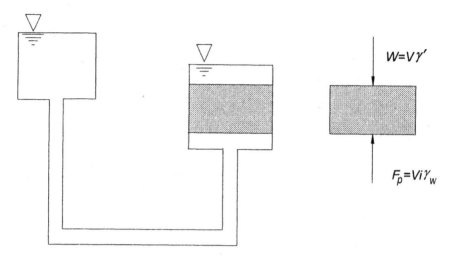

Figure 5.12. Liquefaction condition by an ascending flow

Interpretation of the hydraulic gradient

In the beginning of this chapter the hydraulic gradient was defined as $i = h / L$, which is the ratio between the total hydraulic head h and L, the length of the soil element. Therefore, i represents the energy loss per unit of length of flow. In case of one-dimensional flow in the x direction:

$$i = \frac{\partial \Psi}{\partial x} \text{ or } i = \frac{\partial h}{\partial x} \tag{5.14}$$

Generalizing to three dimensional flow, the hydraulic gradient becomes:

$$i = \frac{\partial h}{\partial x} + \frac{\partial h}{\partial y} + \frac{\partial h}{\partial z} \tag{5.15}$$

Utilizing gradient operator ∇ (this, in fact, was is the origin of the name *hydraulic gradient*), the previous equation can be written as:

$$\vec{i} = \nabla h \tag{5.16}$$

Consequently, Darcy's law (equation 5.3) can be written as:

$$\vec{v} = k \nabla h \tag{5.17}$$

Differential equation of flow

The differential equation of flow is the basis for the study of seepage in soils. This equation assumes validity of Darcy's law; homogeneous soil and water and soil skeleton are incompressible.

The differential equation of flow in an unsaturated soil element, whose deduction is found elsewhere (e.g., Lambe & Whitman, 1979) is:

$$k_x \frac{\partial^2 h}{\partial x^2} + k_y \frac{\partial^2 h}{\partial y^2} + k_z \frac{\partial^2 h}{\partial z^2} = \frac{1}{1+e} \left[S \frac{\partial e}{\partial t} + e \frac{\partial S}{\partial t} \right] \tag{5.18}$$

where: k_j is the permeability in direction j, h is the total hydraulic head (subscript t is avoided for the sake of clarity), S is the degree of saturation, e is the void ratio and t is the time.

Equation 5.18 can be simplified for many applications. Restricting it to two dimensions, saturated soil and steady flow, i.e., $de/dt = 0$ and $dS/dt = 0$, thus:

$$k_x \frac{\partial^2 h}{\partial x^2} + k_y \frac{\partial^2 h}{\partial y^2} = 0 \tag{5.19}$$

Adopting isotropy in relation to permeability, i.e., $k_x = k_y$, the above equation becomes:

$$\frac{\partial^2 h}{\partial x^2} + \frac{\partial^2 h}{\partial y^2} = 0$$

This kind of expression is known as the Laplace equation that governs several physical phenomena such as heat transmission, electric field, etc. It can be written also in a simplified form employing the operator ∇, therefore:

$$\nabla^2 h = 0 \tag{5.20}$$

Equation 5.20 is independent from permeability. As a consequence, its numerical solution will depend only on the geometry of the problem and boundary conditions.

Solution for practical problems of seepage in soils can be obtained in the following ways:

Analytical solutions
Analytical solutions are available for simple cases, usually for isotropic soil, by integration of the Laplace equation. The main references are: Polubarinova-Kochina (1962) and Harr (1962).

The *method of fragments* is a very interesting analytical tool developed in Russia by Pavlovsky (1956), and published in English by Harr (1962), and more recently by Holtz and Kovacs (1981). Analytical solutions are restricted to steady flow, and to constant and isotropic permeability.

Numerical solutions
Numerical solutions are employed for complex situations that can be solved by the use of a computer utilizing the finite difference or the finite element method. Nowadays this is the easiest way to solve a complex problem in non-homogeneous, non-isotropic, unsaturated soil. There are various computer programs available that run on microcomputers. Some have smart graphics packages that allow fast data input and output analysis. Non-saturated soil with varying per-

meabilities and complex two and three dimensional geometries can be easily modeled. The difficulty lies more in the correct evaluation of soil properties rather than in the flow problem.

An important reference on the use of finite differences for solving geotechnical problems is the book of Rushton and Redshaw (1978); some examples of the use of finite element method are dealt with by Veeruijt (1982); both methods are discussed by Franciss (1980).

Electric analogy

The same Laplace equation governs physical problems of electrical, heat and seepage flow. Then, an electric flow model can be used to solve an analogous problem in water flow. A special electrical conducting paper is used to make the model. It is cut in a shape scaled to the prototype. An electrical potential is applied to simulate the boundary conditions of flow. Voltage distribution is measured in several points on the paper and from these data the flow net can be constructed. Details on how to conduct this experiment can be found in Franciss (1980) and Bowles (1970). This analogy has been useful in the past years for many practical applications to solve 2 or even 3 dimensional problems (Cedergren, 1977).

Physical modeling

Sand models in laboratory scale can be employed. Water flow can be traced by the use of dyes and pore pressures can be measured in several points along the model. Andrade (1983) reports practical applications to analyse the complex 3 dimensional drainage system in the foundation of spillway of dams.

Graphical solution

A two-dimensional Laplace equation can be represented graphically through families of curves intercepting each other in right angles forming a *flow net*. This will be discussed next.

Flow net

A *flow net* is a sketch that represents the flow in porous media. It consists of a set of flow and equipotential lines crossing each other in right angles. The flow net can be worked out graphically by a trial and error process described in detail by Cedergren (1977). Numerical methods, such as those mentioned before, and the electrical analogy method can also be used for generating a flow net. Once it is drawn, the determination of pore pressures, hydraulic gradients and discharges is straightforward.

It is out of the scope of this text a comprehensive discussion of the techniques to draw a flow net, as it can be found in Cedergren (1977). Our aim is only to train the reader in the way to use an existing flow net and show him or her how to obtain pore pressures, discharges and gradients. This will be introduced through a simple problem of one dimensional flow presented in figure 5.13. It consists of a sand sample 5 m in height and 2 m x 2 m cross section with a permeability of 5 x 10^{-4} m/s, subjected to a descending flow. The head diagram is presented, including the piezometric, altimetric and total heads. The flow net is shown, consisting of: flow and equipotential lines forming the elements of the net.

Flow lines

Flow lines are lines indicating flow direction. In this problem 5 vertical lines ($n_{fl} = 5$) have been drawn, delimiting four *flow channels* ($n_c = 4$) between two adjacent flow lines.
Then:

$$n_c = n_{fl} - 1 \qquad (5.21)$$

Equipotential lines

Equipotential lines are lines that cross the flow lines at right angles, being the loci of points that have the same potential, or the same energy. In this example, the equipotential lines are vertical. The number of equipotentials, is $n_{eq} = 11$. Between two adjacent equipotential lines there is certain amount of energy loss (ΔH), which can be calculated dividing the total head to be dissipated H by the number of steps n_s between the equipotential lines:

$$\Delta H = \frac{H}{n_s}$$

where:

$$n_s = n_{eq} - 1 \qquad (5.22)$$

Flow net elements

The flow net is composed by rectangular figures or *elements* formed by the interception of equipotentials and flow lines. The length of an element, defined in the direction of the flow, is *l*, and the width, *b*. This example shows only square elements, which is a consequence of the assumption of equal permeability in both vertical and horizontal directions.

Once the geometrical characteristics of the flow net have been defined, calculations of discharge, heads and gradients can be easily performed:

Discharge

The *discharge* or *flow rate* Q_l per unit of length of a flow net is given by:

$$Q_l = k \; H \; \frac{n_c}{n_s} \qquad (5.23)$$

Q_l is the discharge per unit of length, with units of m³/s/m. Note that Q_1 is different from the total flow Q, with units of m³/s. k is the permeability in m/s, and the ratio n_c / n_s is named shape factor, H is the total head loss in metres. This equation will not be deduced here. Instead it will be shown in exercise 5.8, that its results are the same as those obtained by Darcy's law.

Exercise 5.8

Obtain the discharge through the flow net of figure 5.13 using equation 5.23 and compare with the results obtained through Darcy's law.

Solution

From figure 5.13 it comes: $k = 5$ x 10^{-4} m/s, $n_c / n_s = 4/10 = 0.4$ and $H = 8$ m, then:

$$Q_1 = 5 \text{ x } 10^{-4} \text{ x } 0.4 \text{ x } 8 = 1.6 \text{ x } 10^{-3} \text{ m}^3/\text{s/m}$$

As the cross section of the sample in figure 5.13 is 2 m x 2 m, total discharge Q is twice the above value:

$$Q = 3.2 \text{ x } 10^{-3} \text{ m}^3/\text{s}.$$

Applying Darcy's law: $v = Q/A = k \, i$, $\therefore Q = A \, k \, i$. The cross section of the sample is $A = 2$ x $2 = 4$ m², the gradient $i = H/L = 8/5 = 1.6$. Therefore:

$$Q = 4 \text{ x } 5 \text{ x } 10^{-4} \text{ x } 1.6 = 3.2 \text{ x } 10^{-3} \text{ m}^3/\text{s}$$

Heads

The flow net allows the calculation of the total head in any internal point. As a consequence, the piezometric head, and thus the pore pressure, can also be obtained at any point, as shown in exercise 5.9.

Exercise 5.9

Obtain the pore pressure in a piezometer installed at elevation 3 m in the sample of figure 5.13.

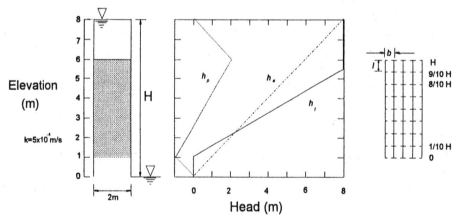

Figure 5.13. Exercise 5.9: one dimensional flow net

Solution

In order to determine the total head at the location of the piezometer, the following steps are followed:

(a) The total head in the upper water level is 8 m.

(b) The head loss in each step or between two adjacent equipotentials is the total head to be dissipated ($H=8$ m) divided by the number of steps ($n_s =10$), therefore: $\Delta H = H/n_s = 8/10 = 0.8$ m.

(c) The total head h_t at the piezometer is obtained by considering the total head in the flow net entry (8 m) minus the accumulated steps up to the equipotential line closest to the location of the piezometer: h_t = 8 m - 6 steps x 0.8 m = 3.2 m.

(d) Piezometric head h at the piezometer location is obtained through equation $h_t = h_p + h_a = 3.2$ m - 3 m = 0.2 m.
    The pore pressure is $u = 0.2$ x $\gamma_w = 0.2$ x $10 = 2$ kPa.

Hydraulic gradients

Hydraulic gradients can be determined in any element in the flow net through:

$$i = \frac{\Delta H}{l} \tag{5.24}$$

where ΔH is the head loss in the element and l is the length in the direction of the flow.

Exercise 5.10

Obtain the gradient i for any element of the flow net of figure 5.13.

Solution

In this one dimensional flow net all elements have the same gradient. Applying equation 5.24 and taking $l = 0.5$ m (measured directly in the flow net) and $\Delta H = 0.8$ m, calculated before (exercise 5.9):

$$i = 0.8/0.5 = 1.6$$

Two dimensional flow

Most of practical problems in geotechnics can be analysed in two dimensions. The flow net in this case presents curved flow and equipotential lines. The following exercises show how to obtain discharges, heads and gradients for a given flow net.

Exercise 5.11

For the sheet pile wall of figure 5.14, obtain the water pressure diagram, the discharge and the exit gradient. Take soil permeability as 3 x 10^{-7} m/s.

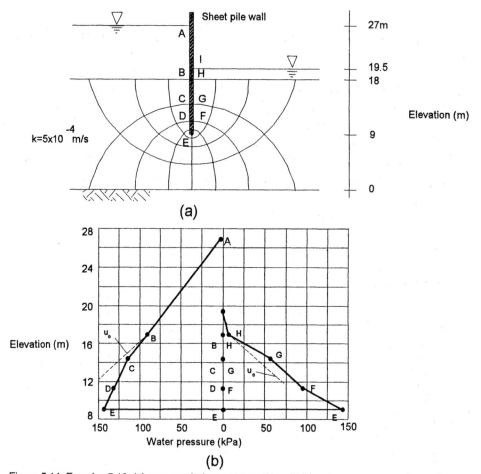

Figure 5.14. Exercise 5.10: (a) seepage below a sheet-pile wall; (b) water pressures on the wall

Solution

(a) The characteristics of the flow net are:

Flow net characteristic	Value
Total head at entry (upstream elevation El_{up})	27 m
Total head at exit (downstream elevation El_{down})	19.5 m
Total head loss $H = El_{up} - El_{down}$	27 - 19.5 = 7.5 m
Number of steps	$n_s = 8$
Number of flow channels	$n_c = 4$
Number of equipotentials	$n_{eq} = 9$
Number of flow lines	$n_{fl} = 5$
Shape factor	$n_c / n_s = 0.5$
Head loss between two adjacent equipotential lines	$\Delta H = 7.5/n_s = 7.5/8 = 0.94$ m

(b) The following table presents data for selected points along the internal and external face of the sheet pile wall. Altimetric head h_a is the elevation of those points, total head h_t is obtained through the equipotentials, the pore pressure obtained from equation 5.7.

POINT	h_a (m)	h_t (m)	h_p (m)	u (kPa)
A	27.00	27.0	0.0	0
B	18.00	27.0	9.0	90
C	14.70	26.1	11.4	114
D	11.70	25.1	13.4	134
E	9.00	23.2	14.2	142
F	11.70	21.4	9.7	97
G	14.70	20.4	5.7	57
H	18.00	19.5	1.5	15
I	19.50	19.5	0.0	0

(c) The discharge is obtained through equation 5.23:

$Q_1 = 3 \times 10^{-7}$ m/s x 7.5 m x 0.5 = 1.13×10^{-6} m³/s/m

(d) The exit gradient is obtained at the exit of the flow net. One should choose the exit element that yields the biggest possible value for exit gradient, since this corresponds to the worst case. In this particular problem, points G and H were chosen.

Previously in this chapter, we discussed the effect of a very high gradient that leads to liquefaction in sands. In addition, high exit gradients may cause *internal erosion*, also called *piping*. This has been the cause of failures of many structures. Such as Teton dam, in the USA (ENR, 1977) and Pampulha dam, in Brazil (Nunes, 1971). Consequently, seepage analysis is necessary in most important structures. The minimum factor of safety to be used is such case is 3, therefore, the highest exit gradient shall not exceed 0.3.

On the other hand, *internal* gradients do not cause problems, even if they present a high value. However, a low gradient is always the safest option. A comprehensive discussion of this matter can be found in Cedergren (1977).

Therefore, the exit gradient for the flow net of figure 5.14 (points G and H) is:

$i = (\Delta H/n_s)/l = (7.5 \text{ m}/8)/3.3 \text{ m} = 0.28$

which is acceptable, since it is below 0.3.

Exercise 5.12

The flow net in the foundation of a gravity spillway of a dam is shown in figure 5.15. It has two sheet pile walls for seepage reduction in the foundation. Cal-

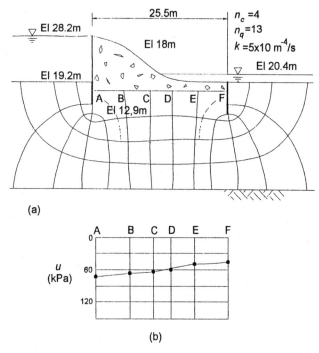

(a)

(b)

Figure 5.15. Exercise 5.12: (a) seepage in the foundation of a gravity dam; (b) Uplift pressures diagram

culate and plot water uplift pressures at the base of the dam, the discharge and the exit gradient. Assume isotropic permeability equal to 5×10^{-5} m/s.

Solution
Following the same steps of the previous exercise.
(a) The characteristics of the flow net are:

Flow net characteristic	Value
Total head at entry (upstream elevation El_{up})	28.2 m
Total head at exit (downstream elevation El_{down})	20.4 m
Total head loss $H = El_{up} - El_{down}$	28.2 - 20.4= 7.8 m
Number of steps	$n_s = 13$
Number of flow channels	$n_c = 4$
Number of equipotentials	$n_{eq} = 14$
Number of flow lines	$n_{fl} = 5$
Shape factor	$n_c / n_s = 0.31$
Head loss between two adjacent equipotential lines	$\Delta H = 7.8/n_s = 7.8/13 = 0.6$ m

(b) Uplift pressures are calculated in the following table:

Point	h_a (m)	h_t (m)	h_p (m)	u (kPa)
A	18.00	25.5	7.5	75
B	18.00	25.2	7.2	72
C	18.00	24.6	6.6	66
D	18.00	24.0	6.0	60
E	18.00	23.4	5.4	54
F	18.00	23.1	5.1	51

(c) The discharge is:

$Q_1 = 5 \times 10^{-9}$ m/s $\times$ 7.8 m $\times$ 4 / 13 = 1.20 $\times 10^{-8}$ m³/s/m

(d) The exit gradient is calculated for the smaller element at the exit of the flow net:

$i = (\Delta H = H / n_s)/ l = (7.8\ m / 13) / 3.5\ m = 0.11$

therefore, i is lower than the 0.3. Therefore, it is acceptable.

Exercise 5.13

The flow net for a homogeneous dam is shown in figure 5.16. The embankment is provided with a downstream slope filter in order to prevent piping. Three Casagrande piezometers are to be installed at points P_1, P_2 and P_3 for pore pressure monitoring. Soil permeability is 2 x 10⁻⁸ m/s. Predict the piezometer readings and obtain the gradient at element X.

Solution
The steps of the previous exercises will be followed.
(a) Flow net characteristics: This is a case of unconfined flow, in which the flow net presents a free or phreatic surface, corresponding to upper flow line or to the water level within the embankment. Additional flow net parameters are summarized below:

Flow net characteristic	Value
Total head at entry (upstream elevation El_{up})	12 m
Total head at exit (downstream elevation El_{down})	0 m
Total head loss $H = El_{up} - El_{down}$	12 - 0= 12 m
Number of steps	$n_s = 8$
Number of flow channels	$n_c = 3$
Number of equipotentials	$n_{eq} = 9$
Number of flow lines	$n_{fl} = 4$
Shape factor	$n_c / n_s = 0.38$
Head loss between two adjacent equipotential lines	$\Delta H = 12/n_s = 12/8 = 1.5$ m

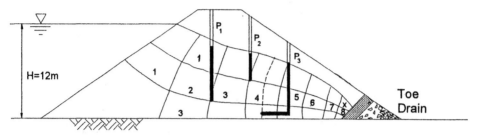

Figure 5.16. Exercise 5.13: seepage through a homogeneous dam

(b) Piezometers readings: Pore pressure predictions at the piezometer locations can be performed as before, by computing total, altimetric and piezometric heads. Alternatively, since this case is an unconfined flow, the altimetric heads can be determined graphically, as shown in figure 5.16. The steps are: first, locate the equipotential line close to the instrument tip where pore pressures are required. Then, interpolate, in case the instrument position does not coincide with the closest equipotential line. Second, follow the chosen equipotential line until it crosses the phreatic surface. The elevation of this point coincides with the piezometric head; in other words, this is the elevation of the top of the water column in the piezometer access tube.

(c) The discharge is calculated as follows:

$$Q_1 = 2 \times 10^{-8} \text{ m/s} \times 12 \text{ m} \times 3/8 = 9 \times 10^{-8} \text{ m}^3 \text{ /s/m}$$

(d) The gradient at X is: taking the length along the flow $l = 1.5$ m at the element X:

$$i = (\Delta H = H / n_s)/l = (12 / 8) / 1.5 \text{ m} = 1$$

which is above the safe limit of 0.3. However, since the dam is only 12 m high, the designer accepted this internal gradient value, but used a downstream slope filter to prevent piping. Modern dams employ other types of internal filters, like the chimney filter.

Proposed exercises

5.1. What is liquefaction due to seepage and how does it occurs?

5.2. Define flow net, flow and equipotential lines.

5.3. What range of permeability should a sand have in order to be regarded as free draining?

5.4. A flow net in the foundation of a concrete dam is shown in figure 5.17. Obtain: (a) The pore pressure at points A, B, C and D. (b) The discharge through the foundation. (c) The hydraulic gradient in element X. Given: $k = 2 \times 10^{-6}$ m/s, $h_1 = 50$ m, $h_2 = 10$ m, $\Delta H = 26$ m, $L = 85$ m.

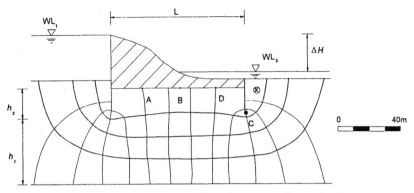

Figure 5.17. Proposed exercise 5.4: seepage through the foundation of a gravity dam

5.5. Why does the foundation treatment in concrete dams involve an upstream grout curtain followed by a downstream drainage well curtains?

5.6. Obtain the discharge through the dam shown in figure 5.18 and the pore pressure at point *P*. Given $k = 2.5 \times 10$ m/s and *H*=18 m.

5.7. Obtain the discharge in the foundation of the dam shown in figure 5.19 and plot the diagram of uplift water pressures and compute the resultant uplift force. Given: $k = 2 \times 10^{-5}$ m/s, $H = 10$ m, $h_1 = 2.8$ m, $h_2 = 1.6$ m, $h_3 = 2$ m.

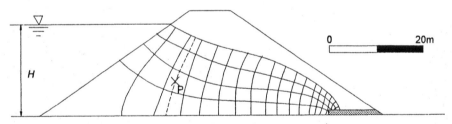

Figure 5.18. Proposed exercise 5.6: seepage through a homogeneous dam

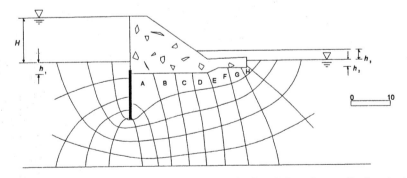

Figure 5.19. Proposed exercise 5.7: seepage through the foundation of a gravity dam having an upstream sheet-pile wall

5.8. For the soil profile in figure 5.20 obtain: (a) σ_v, σ'_v and u in the centre of the silt layer. (b) the flow velocity in the silt layer.

5.9. Obtain the diagram of heads (total, altimetric and piezometric) and the seepage force for the soil sample in figure 5.21.

5.10. Obtain the seepage force acting on the sample in figure 5.22.

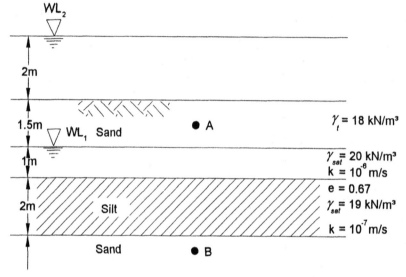

Figure 5.20. Proposed exercise 5.8

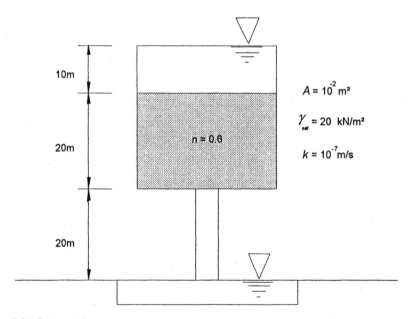

Figure 5.21. Proposed exercise 5.9

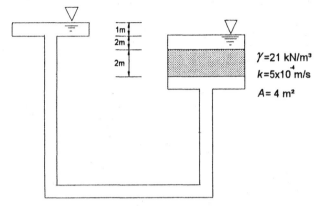

Figure 5.22. Proposed exercise 5.10

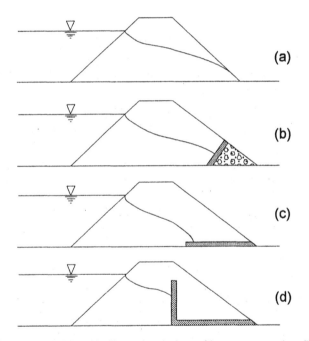

Figure 5.23. Proposed exercise 5.11: Alternative designs of homogeneous dam filters: (a) no filter, phreatic line intercepting downstream slope; (b) toe filter; (c) blanket filter; (d) vertical chimney filter connected with a horizontal blanket filter

5.11. Figure 5.23 presents 4 alternative designs of internal drainage for a homo-geneous embankment dam: (a) no filter, phreatic line crossing the down-stream slope; (b) downstream slope filter; (c) horizontal blanket filter; (d) vertical chimney filter draining to a horizontal filter. Why is type (d) recommended but type (a) is not? Discuss all types. You may consult the recommended references before answering this question.

Chapter 6

Compressibility and settlements

Introduction

This chapter deals with soil compressibility and settlements in conditions of no significant lateral deformation. As an example, consider the embankment foundation in figure 6.1. It has width B and depth D of the compressible layer and $B \gg D$. Points A and B in the foundation present the following behaviour. Point A is below the centre of the loaded area, where there is no shear stress. Volumetric strains at this location will occur without significant change in shape and with no lateral deformation.

Point B lies under the edge of the loaded area and is subjected to significant shear stresses and deformation. As loading is applied, the soil in its vicinity tends to spread out.

This chapter focuses on the behaviour of point A subject to vertical deformation only.

Oedometer test

In order to look at the volumetric strains undertaken by a soil sample, an equipment developed by Terzaghi known as the *oedometer* (*oedos*, from Greek, meaning confined) will be utilized (figure 6.2). A cylindrical soil specimen is confined by a metal ring. Porous stones are employed on the top and bottom to permit drainage of water into or out of the specimen. The vertical load is transmitted by a rigid plate to spread the load. If the specimen is originally saturated in situ, it is kept underwater during the test in order to prevent drying.

The test consists of applying load increments and to observe deformation using a displacement gauge. The imposed conditions are shown in figure 6.3.

Taking solid grains as incompressible, volume change occurs due to the expulsion of gases and pore water, as shown in figure 6.4. It is, then, possible to relate the change in the void ratio of specimen Δe with volumetric strain ε_{vol}.

100

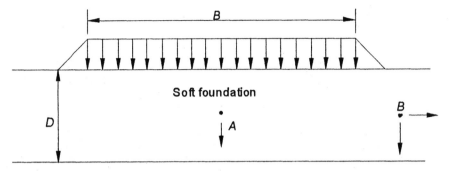

Figure 6.1. Horizontal and vertical displacements under the toe and centre of an embankment during construction

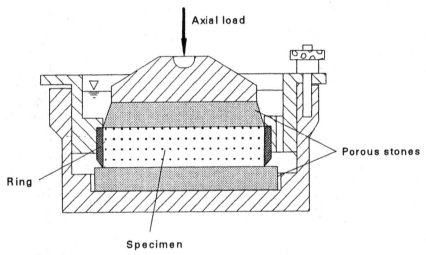

Figure 6.2. The oedometer

In fact, taking:

e_0 = the initial void ratio;
e_1 = void ratio after the deformation;
V_0 = initial volume of the specimen;
V_s = volume of solid particles;
V_v = volume of voids = $e_0 V_s$
V_1 = volume after deformation = $e_1 V_s$

then:

$$\frac{\Delta V}{V_0} = \frac{V_0 - V_1}{V_0} = \frac{V_s(1+e_0) - V_s(1+e_1)}{V_s(1+e_0)} = \frac{e_0 - e_1}{1+e_0}$$

$$\therefore \varepsilon_{vol} = \frac{\Delta V}{V_0} = \frac{\Delta e}{1+e_0} \tag{6.1}$$

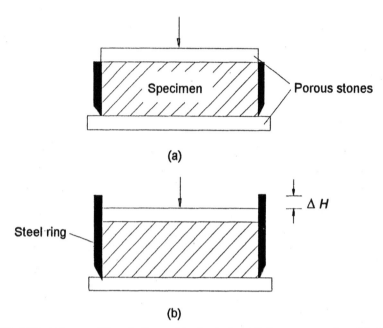

(a)

(b)

Figure 6.3. Deformation conditions in the oedometer test. (a) Before compression; (b) After compression

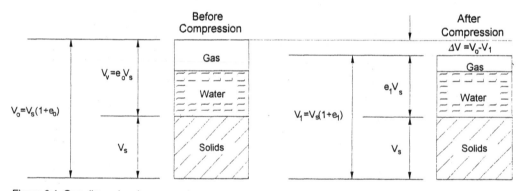

Figure 6.4. One dimensional compression

Behaviour of sands

Sand compressibility in the oedometer will be studied through the test data supplied by Roberts (1964) and Vesic and Clough (1968), included in figures 6.5 and 6.6. In these plots, the abscissae represent the effective vertical stress on the specimen in a logarithmic scale, and the ordinates, the void ratio.

Data in figure 6.5 show negligible volume change or vertical deformation for pressures up to 10 MPa. Only beyond this limit does the specimen begin to deform. This figure also includes data on other granular materials such as: ground quartz and feldspar, which show a similar behaviour to that of the sand sample.

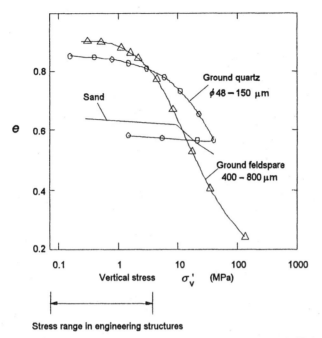

Figure 6.5. Test results of compressibility of sands and other granular materials (Roberts, 1964)

Therefore, there is a limit value of effective vertical stress (σ'_{vm}), beyond which volumetric strain becomes significant with a change in the logarithmic of vertical effective stress σ'_v.

The change in the behaviour of sands before and after σ'_v can be explained by grain crushing. Several researchers (e.g., Roberts, 1964, and Vesic and Clough, 1968) analysed grain size distribution and inspected grain shape in a microscope before and after the test. A change indicates whether or not the grains were crushed. They concluded that there was a threshold value σ'_{vm} for a sand beyond which grains started to crush.

Experiences demonstrate that the particular σ'_{vm} value for a sand is related to the hardness of its grains, or *grain crushability* (Datta et al., 1980, Almeida et al. 1987). In silica or quartz sand, which constitute the majority of sand deposits, σ'_{vm} is usually 10 MPa or over. This pressure is well beyond what is normally applied at ground level by a foundation or an embankment. Therefore, settlements due to sand compressibility are normally neglected in foundation design.

Data in figure 6.6 compare volume change behaviour in loose and dense specimens. They were tested in an isotropic compression cell, enabling high pressures to be applied. The results show that the compressibility is very little influenced by sand relative density, as opposite to σ'_{vm}.

Sand compressibility becomes important when dealing with a material in which the grains are sufficiently weak to allow crushing to occur at low pres-

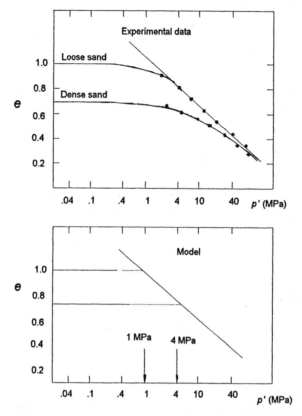

Figure 6.6. Test results on the compressibility of sands (Vesic and Clough, 1968)

sures. This is the case of calcareous sand deposits that occur offshore, as discussed in chapter 1 (figure 1.16). Fortunately, this is seldom the case for most sand deposits on land.

The behaviour of clays

The compressibility of clays will be studied based on data of the Rio de Janeiro soft clay, typical of most soft clay deposits. This material has been investigated for several years at FURJ (e.g., Ortigao and Almeida, 1988).

Oedometer test results from a sample of Rio de Janeiro clay at 5.5 m depth are presented in table 6.1. The test employed six loading steps ranging from 4 to 160 kPa, and then unloaded in 3 steps. Each loading step lasted 24 h. Deformation readings were taken throughout the test. The readings shown in table 6.1 refer to those taken at the end of each step.

Data from table 6.1 were initially plotted in arithmetic scale, as shown in figure 6.7a. The resulting curve is significantly non linear. Two important parame-

Table 6.1. Oedometer test results for Rio de Janeiro soft clay

Phase	σ'_v kPa	ε_v %	e
Loading	0	0	3.60
	4	0.6	3.57
	10	1.8	3.52
	20	3.6	3.43
	40	8.6	3.20
	80	22.1	2.58
	160	33.7	2.05
Unloading	80	32.8	2.09
	10	27.3	2.34
	2.5	24.6	2.47

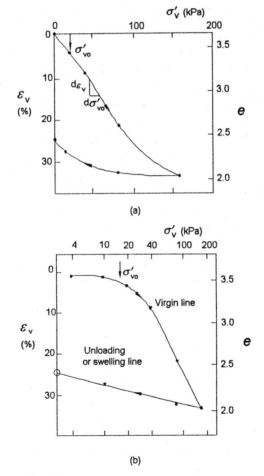

(a)

(b)

Figure 6.7. Oedometer test results of Rio de Janeiro clay

ters are obtained from this plot: the *constrained* or *oedometric modulus* E'_{oed} or M, and its inverse, the *compression modulus*, m_v, defined by:

$$M = \frac{d\sigma'_v}{d\varepsilon_v} \tag{6.2}$$

$$m_v = \frac{d\varepsilon_v}{d\sigma'_v} = \frac{1}{M} \tag{6.3}$$

The constrained modulus is frequently called the *Janbu modulus*, after the Norwegian professor N. Janbu (1963).

The non-linearity of the curve in figure 6.7a, led Terzaghi to represent effective stresses σ'_v in a logarithmic scale, as shown in figure 6.7b. The resulting curve presents a middle part close to a straight line. Terzaghi used this feature as a basis for a model for settlement calculation. This curve can be divided as follows: the first part, starting at the beginning of the curve, is called *recompression phase*, in which the sample is brought back to in situ stresses. At the end of the recompression, the sample presents a significant curvature and then enters a straight line called *virgin consolidation line*, in which it is subjected to significant deformation. Finally, as loads are removed, the clay sample is subjected to the *unloading* or *swelling* phase, in which volume change is relatively small.

The vertical stress at the point where the sample starts to present large deformation when loaded, at the beginning of the virgin consolidation, is called *pre-consolidation* or *overconsolidation stress*. Symbols σ'_{vm} or σ'_{vp} are used.

Determination of overconsolidation stress value is very important since large soil deformation occurs when its value is exceeded. It can be determined from simple oedometer tests through several empirical methods, described in detail by Leonards (1962). Two of these methods have been chosen to be presented here. The first method to appear in literature was devised by Casagrande (1936). The other, which is even simpler than Casagrande's method and more independent from the operator, has been worked out by the late Pacheco Silva (Silva, 1970), a research engineer from the São Paulo Technological Institute.

Casagrande's method

It is shown in figure 6.8 and encompasses the following steps. First, choose the point at the curve that presents the minimum curvature radius. Then, from this point draw two lines, one tangent, another horizontal and determine the bisector of the angle formed by these lines. Now, extend the virgin consolidation line until it intercepts the aforementioned bisector line. The interception points have coordinates (e_{vm}, σ'_{vm}).

From the results in figure 6.8, σ'_{vm} is 34 kPa.

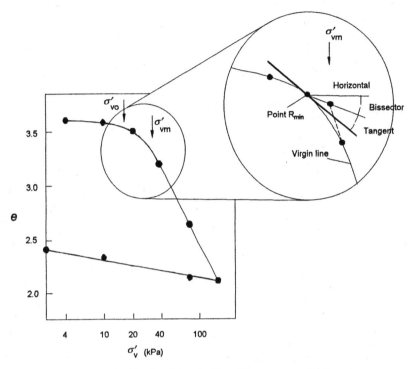

Figure 6.8. Casagrande's method for the determination of the overconsolidation stress

Silva's method

This method is shown in figure 6.9. It encompasses the following steps. First, draw a horizontal line corresponding to ordinate e_0, the initial void ratio. Then, extend the virgin consolidation line and obtain point A of interception with the previous line. From A, draw a vertical line until it intercepts the test curve at B. Then, from B, draw a horizontal line until it intercepts the virgin consolidation line at C, with coordinates (e_{vm} , σ'_{vm}).

From the results in figure 6.9, the value of σ'_{vm} was 30 kPa.

Stress history

Another consolidation test with Rio de Janeiro clay is presented in figure 6.10. It shows the same results as the previous test, but an unload-reload cycle was added, starting at the vertical pressure of 80 kPa. The deformations measured during this cycle were small and reversible and this is a characteristic of an *elastic* behaviour. On the other hand, the virgin consolidation portion of the curve presents characteristics of *plastic* behaviour, since large and irreversible deformation can be observed at this portion of the curve.

The vertical stress of 80 kPa, at the beginning of the unload-reload cycle, rep-

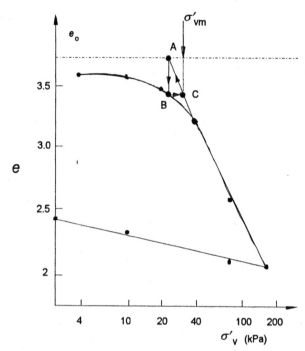

Figure 6.9. Silva's method for the determination of overconsolidation stress

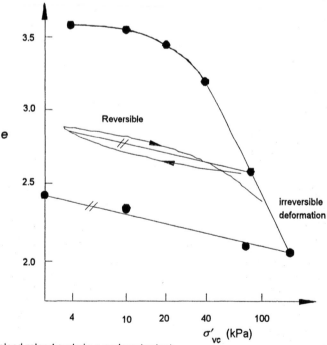

Figure 6.10. Unload-reload cycle in a oedometer test

resents the limit between plastic (virgin consolidation) and elastic behaviour (unloading-reloading). Therefore, once the cycle was unloaded, the clay *remembered* the maximum vertical effective stress $\sigma'_{vm} = 80$ kPa and behaved elastically. At the end of the cycle, the clay was brought back to the virgin consolidation line and started to present large deformation again when the current effective vertical stresses σ'_v became greater than $\sigma'_{vm} = 80$ kPa.

In fact, if it is reloaded back to the virgin consolidation line, it will behave plastically again after suffering stresses beyond $\sigma'_{vm} = 80$ kPa.

It can be concluded that oedometer tests enable to trace the *stress history* of soils, and to obtain the maximum vertical effective stress σ'_{vm}.

According to its stress history, a soil can be classified as normally consolidated (*NC*) or overconsolidated (*OC*) as presented in table 6.2.

The *overconsolidation ratio* (*OCR*) mentioned in table 6.2 is defined as a ratio between the overconsolidation and the current vertical effective stresses:

$$OCR = \frac{\sigma'_{vm}}{\sigma'_v} \tag{6.4}$$

As an example, the in situ *OCR* for the Rio de Janeiro clay at 5.5 m depth is:

$$OCR = \frac{\sigma'_{vm}}{\sigma'_v} = \frac{34}{16} \cong 2$$

Causes of overconsolidation

Consider soil particle *A* (figure 6.11a) during sedimentation. Just after deposition it is submitted to vertical effective stress σ'_{v0}. In this case $\sigma'_{v0} = \sigma'_{vm}$ because the current stress σ'_{v0} has never been exceeded. Therefore, point *A* lies on the virgin consolidation line (figure 6.11b).

Table 6.2. Comparison between current σ'_v and overconsolidation stress σ'_{vm}

Overconsolidated (*OC*)	
$\sigma'_v < \sigma'_{vm}$	• Deformations are small and reversible
	• Elastic behaviour
	• *OCR* > 1
Normally consolidated (*NC*)	
$\sigma'_v \geq \sigma'_{vm}$	• Deformations are large and irreversible
	• Plastic behaviour
	• *OCR* = 1

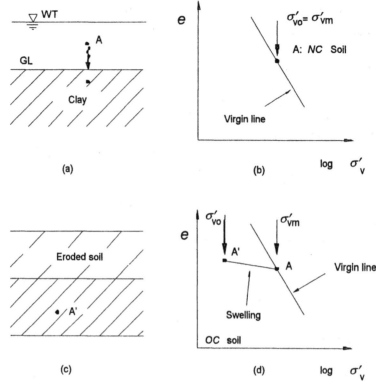

Figure 6.11. (a) and (b) Sedimentation; (c) and (d) Erosion

It is assumed that an erosion process has washed away the top layer of soil above point *A*. The in situ stress will be relieved and the soil element will be unloaded and allowed to swell, as shown in figure 6.11d, becoming overconsolidated.

This is one of the causes of preconsolidation. Other possible causes are: *glaciation*, i.e., the formation and retreat of glaciers in past geological periods. A *change in groundwater level*, which increases the WT height and affects the effective stresses leading to overconsolidation. *Wetting-drying cycles* occurring close to the groundwater table, also may lead to an overconsolidated crust. *Secondary consolidation* or *aging*, as pointed out by Bjerrum (1973), a topic to be discussed in the next chapter. *Leaching*, i.e., precipitation of chemical elements such as silica, aluminum and carbonate compounds of the upper layers in the bottom layers due to rain water infiltration. These phenomena, according to Vargas (1977), are the cause of cementation of grains and the overconsolidated behaviour of the porous São Paulo clay and other clay from the southern regions of Brazil. *Virtual pre-consolidation*, which may occur in residual soils, related, as pointed out by Vargas (1953), to intergranular bonds due to rock weathering. As an example, figure 6.12 shows a residual soil profile near Belo Horizonte,

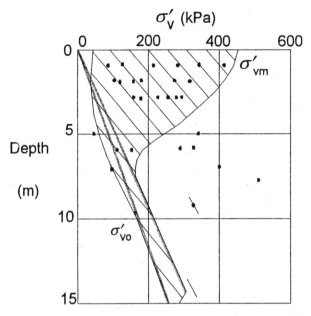

Figure 6.12. Virtual overconsolidation in residual soil from Belo Horizonte, MG, Brazil (Vargas, 1953)

where a strong overconsolidated behaviour was observed in the first 5 m of depth.

Irrespective of the cause, one should always bear in mind that the overconsolidated soil presents small and recoverable (elastic) deformation if stressed below σ'_{vm}. Once σ'_{vm} is exceed the deformations will be large and irreversible, which characterizes a plastic behaviour.

Compressibility parameters

Compressibility parameters represent the inclination of virgin and swelling (or recompression) lines in the oedometer $e \times \log\sigma'_v$ and $\varepsilon_v \times \log\sigma'_v$ plots, as shown in figures 6.13a and b. Table 6.3 summarizes the equations that define these parameters.

Parameters from the $e \times \log\sigma'_v$ plot can be converted into those of the $\varepsilon_v \times \log\sigma'_v$ plot through the following equations:

$$CR = \frac{C_c}{1+e_0} \tag{6.5}$$

$$SR = \frac{C_s}{1+e_0} \tag{6.6}$$

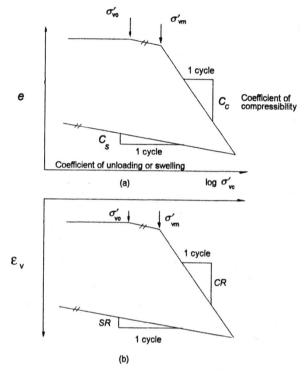

Figure 6.13. Compressibility parameters: (a) $e \times \log \sigma'_v$ curve; (b) $\varepsilon_v \times \log \sigma'_v$ curve

Table 6.3. Compressibility parameters

Plot	Slope	
	Virgin line	**Unloading or swelling line**
	Coefficient of compressibility	Coefficient of unloading or swelling
$e \times \log \sigma'_v$	$C_c = -\dfrac{de}{d\log\sigma'_v}$	$C_s = -\dfrac{de}{d\log\sigma'_v}$
$\varepsilon_v \times \log\sigma'_v$	$CR = -\dfrac{d\varepsilon_v}{d\log\sigma'_v}$	$SR = -\dfrac{d\varepsilon_v}{d\log\sigma'_v}$

Exercise 6.1

Use data from table 6.1 to plot oedometer test curves $e \times \log\sigma'_v$ and $\varepsilon_v \times \log\sigma'_v$ for the Rio de Janeiro clay and obtain the compressibility parameters both graphically and analytically.

Graphical solution

The compressibility parameters will be obtained as shown in figure 6.15:

1. Choose a stress interval corresponding to one log cycle, i.e., between 4 and 40 kPa or 10 to 100 kPa, so that the difference between the logarithms is 1:

   $$\log 40 - \log 4 = 1 \text{ or}$$

   $$\log 100 - \log 10 = 1$$

   For this cycle, C_c can be obtained from:

   $$C_c = -\frac{\Delta e}{\Delta \log \sigma'_v} = \frac{e_{100} - e_{10}}{\log 100 - \log 10} = e_{100} - e_{10}$$

   where: e_{100} and e_{10} are the void ratios at the virgin line corresponding to the chosen log cycle.

2. The virgin line is extended in order to be intercepted by the abscissae corresponding to the selected log cycle.

3. Values of C_c and C_s are obtained graphically are indicated in figure 6.14.

4. An analogous procedure shown in figure 6.15 is used to obtain the parameters CR and SR.

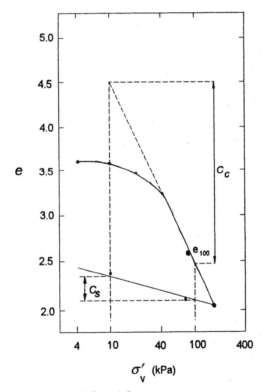

Figure 6.14. Graphical determination of C_c and C_s

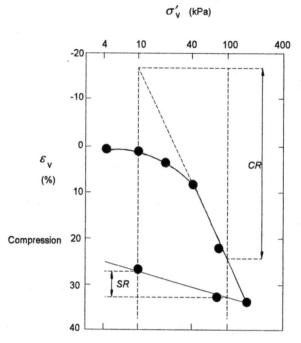

Figure 6.15. Graphical determination of CR and SR

Analytical solution

Applying the equations from table 6.3 for selected points on each portion of the oedometer test curve, it comes:

$$C_c = \frac{3.20 - 2.05}{\log 160 - \log 40} = 1.91$$

$$C_s = \frac{2.47 - 2.05}{\log 160 - \log 2.5} = 0.23$$

$$CR = \frac{33.7 - 8.6}{\log 160 - \log 40} = 0.42 = 42\%$$

$$SR = \frac{33.7 - 24.6}{\log 160 - \log 2.5} = 0.05 = 5\%$$

Alternatively, CR and SR can be obtained from equations 6.5 and 6.6:

$$CR = \frac{1.91}{1 + 3.60} = 0.42 = 42\%$$

$$SR = \frac{0.23}{1 + 3.60} = 0.05 = 5\%$$

Relationship between compressibility and elastic parameters

A relationship between constrained modulus M with Young's modulus E' can be obtained by applying the deformation condition $\varepsilon_2 = \varepsilon_3 = 0$, prevailing in the oedometer, to Hooke's law. Elastic parameters E' and v' refer to *drained* conditions, i.e., drainage and volume change are allowed during deformation. This leads to:

$$M = \frac{E'(1-v')}{(1+v')(1-2v')} \tag{6.7}$$

$$K_0 = \frac{\sigma'_{h0}}{\sigma'_{v0}} = \frac{v'}{1-v'} \tag{6.8}$$

These equations are only valid in the elastic domain, which limits its application to the unloading-reloading phase of the test.

Experience shows that equation 6.8 leads to very poor results and its use is *not recommended*. On the other hand, a very accurate and useful empirical correlation for K_0 with index tests will be presented later in this chapter.

A relationship between m_v and C_c can be deduced considering a small increment de of the void ratio:

$$-de = C_c \; d\log\sigma'_v = \frac{C_c}{2.3}d\ln\sigma'_v = \frac{C_c}{2.3}\frac{d\sigma'_v}{\sigma'_v} \tag{6.9}$$

On the other hand, de can be obtained from m_v:

$$m_v = \frac{d\varepsilon_v}{d\sigma'_v} = \frac{-de}{(1+e_0)d\sigma'_v} \tag{6.10}$$

Combining and rearranging equations 6.9 and 6.10:

$$m_v = \frac{C_c}{2.3(1+e_0)\left[\sigma'_v\right]_{average}} \tag{6.11}$$

Settlement calculation

Settlement ρ can be obtained from equation 6.1:

$$\rho = H_0\frac{\Delta e}{1+e_0} \tag{6.12}$$

where: H_0 is the initial thickness of the compressible layer. This equation is valid regardless of the mechanism causing the change in the void ratio and the degree of saturation of the material. The change in void ratio Δe can be obtained directly at the oedometer curve $e \times \log\sigma'_v$, corresponding to the change in the vertical effective stresses between initial value σ'_{vi} and final value σ'_{vf}.

For a normally consolidated clay, Δe from equation 6.12 gives:

$$C_c = -\frac{\Delta e}{\Delta \log \sigma'_v} = \frac{\Delta e}{\log \sigma'_{vi} - \log \sigma'_{vf}} = \frac{\Delta e}{\log \dfrac{\sigma'_{vf}}{\sigma'_{vi}}}$$

$$\therefore \Delta e = C_c \log \frac{\sigma'_{vf}}{\sigma'_{vi}} \tag{6.13}$$

Combining equations 6.12 and 6.13, it comes:

$$\rho = H_0 \frac{C_c}{1+e_0} \log \frac{\sigma'_{vf}}{\sigma'_{vi}} \tag{6.14}$$

CR can be substituted in equation 6.14, resulting in:

$$\rho = H_0 \; CR \; \log \frac{\sigma'_{vf}}{\sigma'_{vi}} \tag{6.15}$$

The advantage of using *CR* instead of C_c becomes clear in equation 6.15, because void ratio e no longer appears, therefore it is one parameter less.

For the overconsolidated soil, *mutatis mutandis* the following equations are obtained:

$$\rho = H_0 \frac{C_s}{1+e_0} \log \frac{\sigma'_{vf}}{\sigma'_{vi}} \tag{6.16}$$

$$\rho = H_0 \; SR \; \log \frac{\sigma'_{vf}}{\sigma'_{vi}} \tag{6.17}$$

For an overconsolidated soil loaded beyond the overconsolidation stress σ'_{vm}, i.e., $\sigma'_{vf} > \sigma'_{vm}$, equations 6.14 and 6.16 can be combined, giving:

$$\rho = H_0 \left[\frac{C_s}{1+e_0} \log \frac{\sigma'_{vm}}{\sigma'_{v0}} + \frac{C_c}{1+e_0} \log \frac{\sigma'_{vf}}{\sigma'_{vm}} \right] \tag{6.18}$$

By analogy:

$$\rho = H_0 \left[SR \log \frac{\sigma'_{vm}}{\sigma'_{v0}} + CR \log \frac{\sigma'_{vf}}{\sigma'_{vm}} \right] \tag{6.19}$$

Figure 6.16 presents a summary of the equations to be used in the following cases:

(a) Overconsolidated soil loaded beyond the overconsolidated stress, i.e., $\sigma'_{vf} > \sigma'_{vm}$ (figure 6.16a);

(b) Overconsolidated soil loaded less than the overconsolidated stress, i.e., $\sigma'_{vf} > \sigma'_{vm}$ (figure 6.16b);

(c) Normally consolidated soil (figure 6.16c).

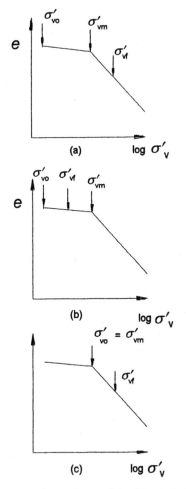

Figure 6.16. (a) *OC* soil loaded with $\sigma'_{vf} > \sigma'_{vm}$; (b) *OC* soil loaded with $\sigma'_{vf} < \sigma'_{vm}$; (c) *NC* soil

Exercise 6.2

Estimate settlements considering the soil profile in figure 6.17, in which an embankment having a height H and unit weight $\gamma = 20$ kN/m³ will be constructed. The value of H can alternatively be 0.5 m, 1 m and 3 m. The geotechnical properties for this clay are: $C_c = 1.91$, $C_s = 0.16$, $e_0 = 3.6$, $\sigma'_{vm} = 34$ kPa, $\gamma = 13$ kN/m³ .

Solution

1. Considering the embankment height $H = 0.5$ m, and assuming the clay is homogeneous, the settlements will be calculated in the middle of the clay layer:

$$\sigma'_{v0} = 5.5 \text{ m x } (13-10) \text{ kN/m}^3 = 17 \text{ kPa}$$

$$\sigma'_{vf} = \sigma'_{v0} + \Delta\sigma = 17 + 0.5 \text{ m x } 20 \text{ kN/m}^3 = 27 \text{ kPa}$$

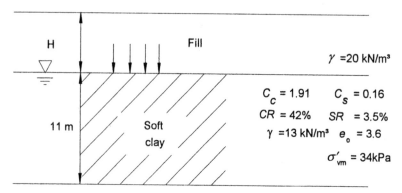

Figure 6.17. Exercise 6.2

$$\therefore \sigma'_{vf} < \sigma'_{vm}$$

Settlement is calculated through equation 6.16:

$$\rho = 11\frac{0.16}{1+3.6}\log\frac{27}{17} = 0.08 \text{ m}$$

2. If the embankment height is $H = 1$ m, then:

$$\sigma'_{vf} = \sigma'_{v0} + \Delta\sigma = 17 + 1 \text{ m} \times 20 \text{ kN/m}^3 = 37 \text{ kPa}$$

$$\therefore \sigma'_{vf} > \sigma'_{vm}$$

Equation 6.18 is applied:

$$\rho = 11\left[\frac{0.16}{1+3.6}\log\frac{34}{17} + \frac{1.91}{1+3.6}\log\frac{37}{34}\right] = 0.28 \text{ m}$$

3. If the embankment height is $H = 3$ m, then:

$$\sigma'_{vf} = \sigma'_{v0} + \Delta\sigma = 17 + 3 \text{ m} \times 20 \text{ kN/m}^3 = 77 \text{ kPa}$$

$$\therefore \sigma'_{vf} > \sigma'_{vm}$$

Equation 6.18 is applied:

$$\rho = 11\left[\frac{0.16}{1+3.6}\log\frac{34}{17} + \frac{1.91}{1+3.6}\log\frac{77}{34}\right] = 1.75 \text{ m}$$

Exercise 6.3

Estimate settlements for the profile of figure 6.17 of Rio de Janeiro clay, where a 2 m high sand embankment will be constructed with unit weight of $\gamma = 18$ kN/m³. Use clay properties from several oedometer tests presented in figure 6.18. The unit weight of the clay is $\gamma = 13$ kN/m³.

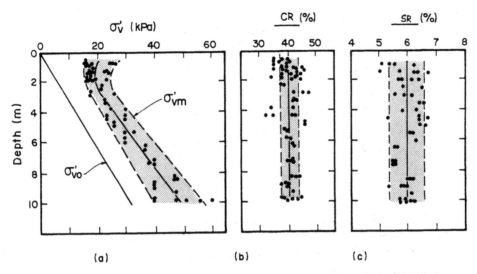

Figure 6.18. Results from oedometer tests on Rio de Janeiro clay. (a) Stress history; (b) Virgin compression ratio; (c) Swelling compression ratio

Solution
As clay properties vary with depth, the clay layer will be divided into sub-layers. Settlements will be computed separately in the middle of each sub-layer, and then summed up in the end. Average values of geotechnical properties from figure 6.18 will be considered. Equation 6.18 is employed in the calculations:

z	H_0	$z_{average}$	σ'_{v0}	σ'_{vm}	σ'_{vf}	CR	SR	ρ
m	m	m	kPa	kPa	kPa	%	%	m
0 - 2	2	1.0	3.0	19	39	40	6	0.10 + 0.25
2 - 5	3	3.5	10.5	23	47	40	6	0.06 + 0.37
5 - 8	3	6.5	19.5	34	56	40	6	0.04 + 0.26
8 - 11	3	9.5	28.5	46	65	40	6	0.04 + 0.18
							$\rho = \sum$	1.3

Correlations

Correlations between compressibility parameters and simple index properties are useful tools in engineering practice. In the preliminary stage of a project a geotechnical engineer may not have all the test data that he or she needs and may utilize a simple correlation between C_c with Atterberg limits for estimating settlements.

Later in the project, after the engineer has received the final site investigation report, such correlations may still be useful to evaluate results. Then, the engineer compares test data with other clays from the same or other sites. When a

Table 6.4. Correlations $C_c = f(LL)$

Site	Correlation	Reference
Chicago, USA	$C_c = 0.01\ w$	Azzouz et al. (1976)
Clays of low sensitivity	$C_c = 0.009\ (LL - 10)$	Terzaghi & Peck (1967)
Tertiary clays of São Paulo, Brazil	$C_c = 0.0046\ (LL - 9)$	Cozzolino (1961)
Marine clay from Santos, Brazil	$C_c = 0.0186\ (LL - 30)$	"
Marine clay from Rio de Janeiro, Brazil	$C_c = 0.013\ (LL - 18)$	Ortigao (1975)

big difference is observed for similar clays with similar origin, the engineer will investigate its possible causes. Common arguing can be: is the clay different or are the results wrong? How does the sampling technique affect the results? How was the testing performed?

Table 6.4 shows some examples of correlations of the type between C_c and the liquid limit LL for several sedimentary soils. These correlations usually present a wide data scatter, in the order of 30%, therefore, they are only locally valid.

In tropical lateritic and saprolitic soils, correlations of C_c with the liquid limit present excessive scatter (Lacerda, 1985, Mililitsky, 1986), thus, it is preferred to correlate C_c with void ratio e. Figure 6.19 shows a relationship for soils of different origins.

A few correlations can be applied to soils of different geological origin, and are, therefore, universal. Among those, are equations 6.20 and 6.21.

$$C_c = \frac{1}{2}\left[\frac{\gamma_w}{\gamma_d}\right]^{\frac{12}{5}} \tag{6.20}$$

$$CR = 0.329\left[1 + \frac{0.0133\,PI(1.192 + A_c^{-1}) - 0.027\,PL - 1}{1 + 0.027w}\right] \tag{6.21}$$

where: γ_w = the unit weight of water 10 is kN/m³; $\gamma_d = \gamma_t / (1 + w)$ the dry unit weight of the clay; A_c the activity $A_c = PI / (\% < 2\mu\,\text{m})$; w = water content (%); PL = the plastic limit (%); LL = the liquid limit (%).

Figure 6.20 presents a chart for the graphical solution of equation 6.21.

Exercise 6.4

Estimate C_c and CR for Rio de Janeiro clay from index properties (figure 1.14) using the following average data: $\gamma = 13$ kN/m³ , $PI = 80\%$, $PL = 40\%$, $w = 150\%$, $e = 3.6$. The percentage of clay (i.e., $\% < 2\ \mu\text{m}$) is 55%.

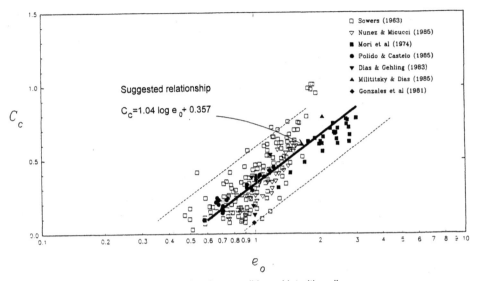

Figure 6.19. Correlation between C_c and e_0 for saprolitic and lateritic soils

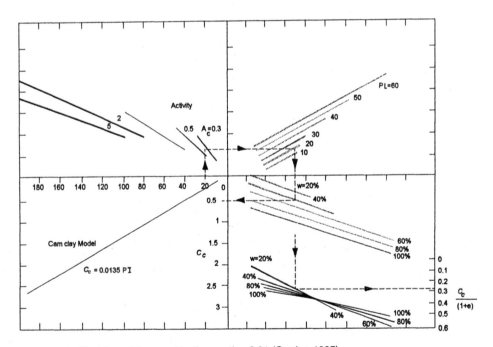

Figure 6.20. Chart for solving graphically equation 6.21 (Carrier, 1985)

Solution

1. Using equation 6.20:

$$\gamma_d = \frac{13}{1 + \frac{150}{100}} = 5.2 \text{ kN/m}^3$$

$$C_c = \frac{1}{2}\left[\frac{10}{5.2}\right]^{\frac{12}{5}} = 2.4$$

2. Using equation 6.21:

$$CR = 0.329\left[1 + \frac{0.0133 \times 80(1.192 + 1.45^{-1}) - 0.027 \times 40 - 1}{1 + 0.027 \times 150}\right] = 0.32 = 32\%$$

K_0 from oedometer tests

In standard oedometer tests horizontal stress σ'_h is not measured, therefore, the coefficient K_0 cannot be obtained during the test. However, if a special horizontal stress transducer is installed, K_0 could be calculated from equation 3.10 for each stage of loading.

In order to look at K_0 value during an oedometer test in clay, data on kaolin (Nadarajah, 1973) shown in figure 6.21 will be used. The specimen was initially normally consolidated at the vertical effective stress σ'_v of 50 kPa, corresponding to the first loading stage (point A in figure 6.21a). Additional load stages were applied until σ'_v reached 550 kPa (point B). Then, the specimen was unloaded to 80 kPa (point D). The line between AB corresponds to a normally consolidated behaviour; BD, to an overconsolidated behaviour.

Figure 6.21b presents the variation of K_0 versus σ'_v during the test. It is noteworthy that K_0 is approximately constant and close to 0.55 during loading, but increases during unloading reaching a value near 1.5 at the end of the test. It is possible to establish a relationship between K_0 and OCR, shown in figure 6.21c. This plot demonstrates that K_0 value has a strong dependency on OCR.

In summary, these data allow to conclude that in normally consolidated clays, K_0 is approximately constant and lies in the 0.5-0.6 range. However, if the clay is overconsolidated, K_0 increases, reaching values greater than 1.

Similar work on sands (e.g., Al Hussaini et al., 1975, Daramola, 1980, Mayne and Kulhawy, 1982) extended these conclusions to granular soils.

Jaky (1944) correlated K_0 in normally consolidated clays with the mobilized friction in soil, obtaining:

$$K_0 = 1 - \sin\phi' \tag{6.22}$$

Parameter ϕ' is the effective friction angle of the soil, which will be studied in chapters 9 and 10. Although totally empirical, Jaky's equation has been found to

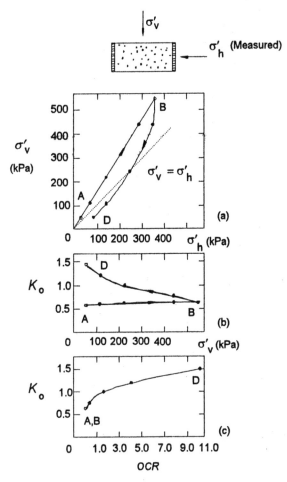

Figure 6.21. Oedometer test in kaolin (Nadarajah, 1973)

give surprisingly good results, for both normally consolidated sands and clays as shown in figure 6.22.

For overconsolidated soils, Mayne and Kulhawy (1982), proposed the following modification to Jaky's formula:

$$K_0 = 1 - \sin\phi' \ OCR^{\sin\phi'} \tag{6.23}$$

Stress paths in oedometer tests

A stress path can only be drawn for a special oedometer test in which the horizontal stress is known, and K_0 values are calculated. Figure 6.23a presents an *ESP* for this special test. *AB* corresponds to the first loading on the clay, which is normally consolidated, and K_0 value is fairly constant. Line *AB* plots on the K_0 *line*.

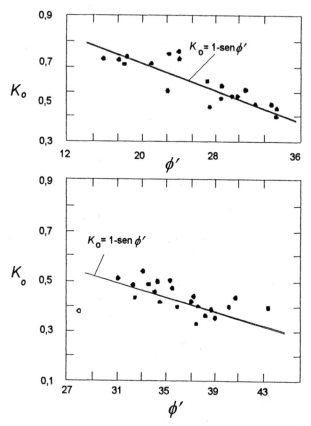

Figure 6.22. K_0 in *NC* soils: (a) clays (Ladd et al., 1977); (b) sands (Al Hussaini et al., 1975)

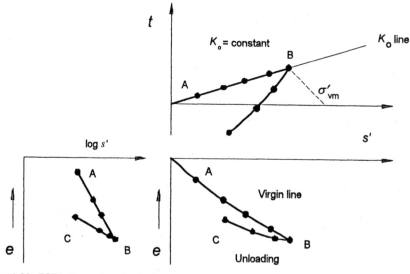

Figure 6.23. *ESP* in the oedometer test

The unloading phase starts at *B* and, thereafter, the clay becomes overconsolidated. The *OCR* value increases as the current σ'_v decreases and K_0 can be estimated through equation 6.23. As K_0 varies with *OCR*, line *BC* cannot be linear.

Figure 6.23b presents the *s':e* diagram that will be frequently used with the *ESP*. The virgin and the swelling lines are curved in this plot. Then, it is convenient to have an auxiliary plot log *s':e* in which both virgin and swelling lines will be straight lines.

The *s':t:e* diagram in figure 6.23 is a very useful tool for the interpretation of both stresses and volumetric strains in a soil element and will be employed throughout this book. A three-dimensional representation for the *s':t:e* diagram, used by several researchers (e.g., Atkinson and Bransby, 1978), will be avoided here for the sake of simplicity.

Equations for the oedometric and isotropic consolidation lines

The slopes of the virgin and the swelling lines in the *e*:log *s'* diagram (figure 6.24) are - C_c and - C_s, respectively. Therefore, considering e_{c0} as the void ratio at the virgin line corresponding to *s'* = 1 kPa, the equation of the virgin consolidation line becomes:

$$e = e_{c0} - C_c \log s' \tag{6.24}$$

By analogy, for isotropic compression:

$$e = e_c - C_c \log s' \tag{6.25}$$

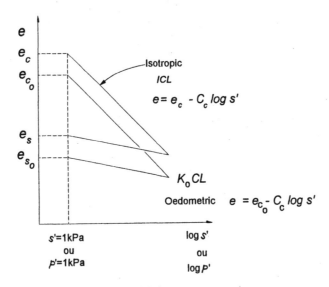

Figure 6.24. Isotropic and oedometric consolidation

The unloading lines become:

oedometric $e = e_{s0} - C_c \log s'$ (6.26)

isotropic $e = e_s - C_c \log s'$ (6.27)

Equations 6.24 to 6.27 can also be defined with p' in place of s', as we shall see later in chapter 14.

Exercise 6.5

Plot the $s':t:e$ diagram for an oedometer compression having the following loading stages:

loading: 80, 300 and 600 kPa

unloading: 300, 150 and 75 kPa.

K_0 values can be obtained through equation 6.23 taking $\phi' = 25°$.
   Given: $C_c = 2.07$, $C_s = 0.28$, the void ratio corresponding to the first loading stage is 2.58.

Solution
Equation 6.23 is used to obtain K_0 values, according to the calculation presented below. Then, σ'_h is obtained through equation 3.10, and s' through $s' = 0.5 (\sigma'_v + \sigma'_h)$.

σ'_v	OCR	K_0	σ'_h	s'	e
kPa			kPa	kPa	
80	1	0.58	46	63	2.58
300	1	0.58	174	237	1.39
600	1	0.58	348	473	0.77
300	2	0.78	234	266	0.84
150	4	1.04	156	153	0.93
75	8	1.39	104	90	1.01

   The void ratios for the virgin and swelling lines (last columns of the table above) were obtained through equations 6.24 and 6.26. The corresponding values of e_{c0} and e_{s0} are calculated starting at the first loading stage, for which the void ratio of 2.58 is given.
   Therefore, entering equation 6.24 with data for the first loading stage: $s' = 63$ kPa and $e = 2.58$:

   $2.58 = e_{c0} - 2.07 \times \log 63$

   $\therefore e_{c0} = 6.3$

The virgin line equation becomes:

   $e = 6.3 - 2.07 \log s'$

This equation enables the calculation of the void ratio for the other compression

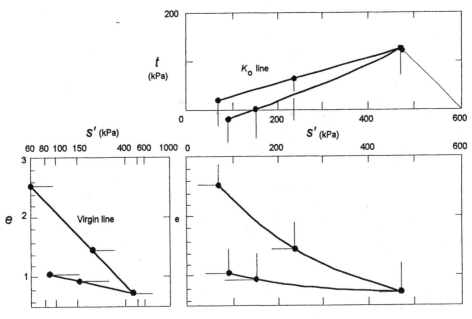

Figure 6.25. Exercise 6.5

steps. Unloading starts at ($s' = 473$ kPa, $e = 0.77$) and an analogous procedure is applied. Therefore, using equation 6.26:

$$0.77 = e_{s0} - 0.28 \times \log 473$$

$$\therefore e_{s0} = 1.5$$

The swelling line is, then:

$$e = 1.5 - 0.28 \log s'$$

enabling the calculation of void ratios for the unloading stages. The resulting $s':t:e$ plot is represented in figure 6.25.

Collapsible soils

Some soil deposits in very dry environments may exhibit excessive volumetric deformation when saturated. This is the case for loess deposits, transported by wind, and clays subjected to leaching. Leaching is a process in which the soluble compounds are dissolved and, then, transported by means of water flow. It takes place in very dry environments following a period of heavy rain. Collapsible soils have been found in many countries, e.g., Australia, Brazil, USA, etc.

Some tropical soils are also collapsible. A few examples of these soils are the porous clays of São Paulo (Vargas, 1973) and Brasília, Brazil. As an example, a summary of soil properties of the porous clay of Brasília is presented in fig-

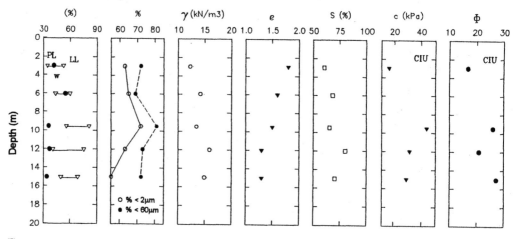

Figure 6.26. Soil properties of the porous clay of Brasília (Ortigao and Macedo, 1993)

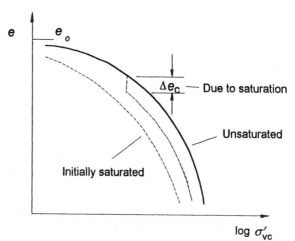

Figure 6.27. Collapse due to saturation (Vargas, 1977)

ure 6.26. It is named *porous* due to the high void ratio, around 1.7, and low unit weight, close to 13 kN/m³, at the top of the layer. This is due to leaching.

The effects of collapsibility can be very severe even on light one-story buildings on shallow foundations. Foundation practice in the regions is to use deep piles to transfer load to a more competent and deep soil layer.

In arid regions small embankment dams may be compacted at a very low water content, below the optimum value, due to the lack of water. At the first reservoir filling, the embankment collapses due to saturation (Miranda, 1988).

The oedometer test can be used to study the effect of saturation, as shown in figure 6.27. The test results of three identical specimens are compared; one was tested fully saturated, another was tested dry, and a third, initially dry, had water added at some loading stage to promote saturation. The third specimen exhibited a change in void ratio equal to Δe_c due to saturation. This corresponds to a volumetric strain $\Delta \varepsilon_c$. The volumetric strain is related to the previous value by the expression $\Delta \varepsilon_c = \Delta e_c (1 + e_0)$, where e_0 is the initial void ratio of the dry specimen. According to Vargas (1973), the soil can be said to be collapsible if $\Delta \varepsilon_c > 2\%$. There are however, other criteria for the characterization of collapsible soils (Vilar et al., 1981), which will not be discussed herein.

Expansive soils

Some tropical soils present an opposite behaviour. They expand in presence of water. This is frequently the cause of accidents in constructions and pavements. These soils occur in arid and semi-arid regions and contain expansive minerals like montmorilonite or illite. After very long periods of drought, the ground water level is very low, several metres below the surface. Sometimes they can be identified because the surface of these deposits presents several fissures in a honeycomb pattern, due to shrinking. When wet by rain, construction activity nearby, or pipeline leakage, they swell, leading to defects in structures.

A full discussion of expansive soils is out of the scope of this book but can be found elsewhere (e.g., Hunt, 1984, Nunes, 1978, and Simmes and Costa-Filho, 1981).

Proposed exercises

6.1. What is overconsolidation stress in sands and clays and what is its practical significance?

6.2. Why do calcareous sands present significant volumetric strain when loaded even in small stresses, as opposite to silica sands?

6.3. Define the following parameters and their equations: σ'_{vm}, M, m_v, OCR, C_c, C_s, CR, SR and K_0.

6.4. Deduce the equation $\varepsilon_{vol} = \Delta e(1 + e_0)$.

6.5. An oedometer test in a clay specimen having $e_0 = 0.965$ presented the following results:

σ'_v (kPa)	e
20	0.953
40	0.948
80	0.938
160	0.920
320	0.878
640	0.789
1280	0.691
320	0.719
80	0.754
20	0.791

Plot the $e \times \log \sigma'_v$ curve and obtain the following parameters σ'_{vm}, *OCR*, C_c, C_s, *CR*, *SR*. Plot the $e \times \sigma'_v$ curve and obtain *M* and m_v variation with the effective stress σ'_v. Considering this test representative of a 10 m thick clay layer in which $\gamma = 13$ kN/m^3, compute settlements due to a loading of 300 kPa applied at the ground level. The water table is at ground surface.

6.6. Estimate settlements due to a 3 m high embankment ($\gamma = 18$ kN/m^3) on Rio de Janeiro clay. Take test data from table 6.1. Use 4 sub-layers in the computations.

6.7. Plot the *s':t:e* diagram for a clay presenting $\phi' = 30°$, $C_c = 0.65$, $C_s = 0.04$ and $e_0 = 5.2$. The clay was initially normally consolidated under $\sigma'_v = 100$ kPa, was then loaded in a oedometer up to $\sigma'_v = 320$ kPa, and unloaded to $\sigma'_v = 20$ kPa. K_0 can be estimated through equation 6.23.

6.8. Redo previous exercise for isotropic compression and take $e_c = 5.7$.

6.9. What does collapse due to saturation mean and what is its importance in soils engineering?

Chapter 7

Consolidation

Introduction

Saturated deposits of low permeability soils, when loaded by a surcharge, may lead to settlements taking place over a long period of time. As an example, structures built on alluvial or marine marshy areas, where soft clay deposits frequently occur, may deform at a very slow rate. This phenomenon is known as consolidation and has firstly been studied by Terzaghi in 1914, when teaching at the University of Constantinople. He developed the oedometer test, a method for settlement calculations and later accounted for time rate in his consolidation theory.

Terzaghi's piston-spring-water system analogy

The piston-spring-water system analogy consisted of a physical model of the consolidation process developed by Terzaghi. Consider a sample of saturated and low permeability clay loaded by an increment of vertical stress $\Delta\sigma_v$ in a oedometer (figure 7.1a). Voids in the soil matrix are fully saturated by water. A pressure gauge is employed to observe pore water pressure.

Figure 7.1b presents Terzaghi's model of consolidation. It consists of a very rigid cylinder filled with water, having a frictionless piston and a valve for control of water flows. The piston is supported by a spring that represents the compressibility of soil skeleton. The valve represents the permeability of soil.

When loaded by vertical stress increment $\Delta\sigma_1$, the water pressure rises, as indicated in the pressure gauge. At the initial time $t = 0$ the valve is still closed and the gauge pressure is equal to the applied stress: $\Delta u_{t=0} = \Delta\sigma_1$. The water pressure supports all the external loads and, since the water is essentially incompressible, the spring is still unloaded.

As the valve is opened and water starts to be expelled, water pressure slowly decreases transferring load to the spring. This represents the process that occurs

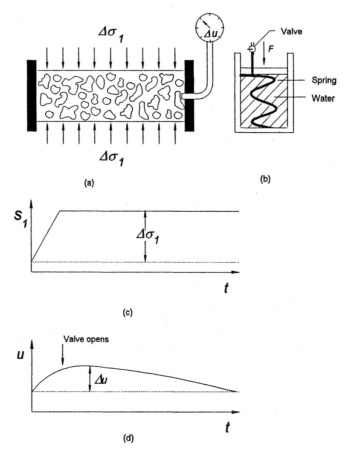

Figure 7.1. (a) Test conditions at the consolidation test; (b) Terzaghi's spring-water system analogy; (c) Loading versus time; (d) Pore pressure versus time

in soils: pore pressures decrease, opposite to the increase of effective stresses in the soil skeleton.

Figures 7.1c and 7.1d present the change in total vertical stress σ_1 and the pore pressure u that occurs with time. Excess pore pressure Δu is the difference between current pore pressure u and equilibrium value u_0. After the valve is opened, dissipation takes place and pore pressure tends to equilibrium, while the effective stress in the soil skeleton increases.

Terzaghi's one dimensional consolidation theory

The theory of consolidation was introduced in 1925 in Terzaghi's book *Erd-baumechanik*. It is important to understand its assumptions and its limitations.

In order to fit a theoretical framework to the piston-spring-water analogy it is necessary:

(i) an equation to represent the water flow;
(ii) another for spring compressibility;
(iii) a third one, to ensure equilibrium.

For the flow of water, the continuity equation of flow, reviewed in chapter 5 (equation 5.16), can be rewritten for the one-dimensional case as:

$$k\frac{\partial^2 h}{\partial z^2} = \frac{1}{1+e}\left[S\frac{\partial e}{\partial t} + e\frac{\partial S}{\partial t}\right] \qquad (7.1)$$

where: k = permeability in the vertical direction; z = coordinate in the vertical direction; h = total hydraulic head; e = void ratio; S = degree of saturation; t = time.

The use of equation 7.1 implies in the following assumptions:

Validity of Darcy's law - The proportionality between flow velocity and the hydraulic gradient has been proved even in very low gradients that occur in clays during consolidation (Tavenas et al., 1983). Therefore, Darcy's law can be extended without restrictions to consolidation.

Small strains - This assumption considers that the thickness of the soil layer under consolidation does not change with time. Resulting strains are so small that they do not affect the initial dimensions of the problem. In fact, the majority of practical problems, even when settlements represent 30% of the initial layer thickness, can be dealt with small strains with little error. There is, however, a class of problems that has to be dealt with in finite or large strains. A typical problem is the modeling of sedimentation process in waste stabilization ponds, in which the material is placed as a sludge. The vertical deformation due to consolidation may reach 90% of the initial thickness and any prediction based on Terzaghi's theory may be erroneous.

Soil particles and water are incompressible - The compressibility of water is very low as compared to the soil skeleton and can be neglected. Solid grains can also be treated as incompressible since all deformation is taken by the soil skeleton, which is represented by spring in Terzaghi's analogy.

One dimensional flow - This hypothesis is valid when the thickness of the soil layer under consolidation is smaller than the breadth of the loaded area (figure 7.2).

Terzaghi's theory also applies the following additional restrictions:

Saturated soil - Considering full saturation: $S = 1$ and $dS/dt = 0$. Equation 7.1 simplifies to:

$$k\frac{\partial^2 h}{\partial z^2} = \frac{1}{1+e}\frac{\partial e}{\partial t} \qquad (7.2)$$

Total head h is the sum of altimetric head h_a and piezometric head h_p. The value of h_p is equal to pore pressure u divided by unit weight of water γ_w (equations 5.6

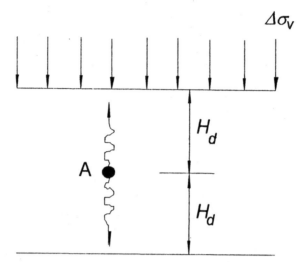

Figure 7.2. One dimensional consolidation and the path of a water particle in the middle of the clay layer

and 5.7), thus:

$$h = h_a + h_p = h_a + \frac{u}{\gamma_w}$$

The value of u can be replaced by $u + \Delta u$. Where u is the equilibrium pore pressure and Δu the out-of-balance increment due to the applied load. Therefore:

$$h = h_a + (u_0 + \Delta u)/\gamma_w \qquad (7.3)$$

Applying the differential operator $\partial^2/\partial z^2$ to equation 7.3, the conclusion is that $\partial^2 h_a/\partial z^2 = 0$ and $\partial^2 u_0/\partial z^2 = 0$. Therefore:

$$\frac{\partial^2 h}{\partial z^2} = \frac{1}{\gamma_w} \frac{\partial^2 \Delta u}{\partial z^2} \qquad (7.4)$$

Writing $\partial \Delta$ is a mathematical heresy because an increment cannot be differentiated. Thus, ∂u will be used instead, where u is the out-of-balance pore pressure.

Consequently, equation 7.2 is rewritten:

$$k \frac{1}{\gamma_w} \frac{\partial^2 u}{\partial z^2} = \frac{1}{1+e} \frac{\partial e}{\partial t} \qquad (7.5)$$

Two additional hypotheses are now introduced, one for the soil skeleton, another for the equilibrium condition:

Linear stress-strain relationship - Terzaghi adopted for the soil skeleton the following linear stress-strain relationship:

$$\frac{\partial e}{\partial \sigma'_v} = -a_v \qquad (7.6)$$

where: σ'_v is the vertical effective stress and a_v a *one-dimensional compression modulus*. Combining equations 7.6 in 7.5 and rearranging terms:

$$k\frac{(1+e)\partial^2 u}{\gamma_w a_v \partial z^2} = -\frac{\partial \sigma'_v}{\partial t} \qquad (7.7)$$

The independent term to the left of equation 7.7 was called by Terzaghi the *coefficient of consolidation* c_v:

$$c_v = \frac{k(1+e)}{\gamma_w a_v} \qquad (7.8)$$

In equation 7.8 the relationship $(1+e)/a_v$ is the inverse of the modulus m_v defined before in chapter 6 (equation 6.3). Therefore:

$$c_v = \frac{k}{\gamma_w m_v} = \frac{kM}{\gamma_w} \qquad (7.9)$$

c_v *remains constant during consolidation* - This simplifying assumption by Terzaghi does not match real soil behaviour, since the coefficient of consolidation is not an independent property, but varies with the permeability and compressibility of the soil. Both parameters decrease as consolidation proceeds, leading to a net decrease of c_v. Laboratory and field evidence show that the coefficient of consolidation is high for overconsolidated clays, being considerably reduced when the clay becomes normally consolidated. Therefore, the assumption of constant c_v is at least a very rough approximation.

Equation 7.7 is, then, rewritten as:

$$c_v\frac{\partial^2 u}{\partial z^2} = -\frac{\partial \sigma'_v}{\partial t} \qquad (7.10)$$

Equilibrium condition - Terzaghi's theory assumes that total stresses do not vary during consolidation:

$$\sigma_v = \sigma_{v0} + \Delta\sigma_v = \text{constant}$$

where: σ_v is the total vertical stress, σ_{v0} is the initial total vertical stress, and $\Delta\sigma_v$ the applied stress increment, assumed constant in the soil layer. This hypothesis implies that excess pore pressure Δu corresponds to an equivalent change in effective stress σ'_v, i.e., $\partial u = -\partial \sigma'_v$.

Terzaghi's one dimensional differential equation

Entering the above assumption into equation 7.10 leads to Terzaghi's one dimensional differential equation:

$$c_v\frac{\partial^2 u}{\partial z^2} = \frac{\partial u}{\partial t} \qquad (7.11)$$

This is a partial derivatives equation of the second order that can be solved by exact closed form solutions or numerical approximations.

Exact solution of the consolidation equation

A closed form solution for equation 7.11 was obtained by Terzaghi in his book *Erdbaumechanik*. He adopted the following boundary and simplifying conditions:

Unit weight of clay is neglected - The consolidation under self weight, like the sedimentation process, cannot be analysed by Terzaghi's original solution. The assumption is valid, however, to analyse the effect of a surcharge applied at the ground level.

Isotropic behaviour - This means adopting initial pore pressure increments equal to the surcharge load, i.e., $\Delta u_{t=0} = \Delta \sigma_v$. This is only valid when the breadth of loading is bigger than the thickness of the clay layer.

Drainage at the top and bottom of the clay layer - If a clay layer under consolidation presents drainage at the top and at the bottom, it has *double drainage*. This is very common in practice. Terzaghi's solution can also be easily extended to cases of single drainage. Drainage assumptions play a major role in the result. In large projects it should be checked by field monitoring with piezometers.

The mathematical solution for this case is deduced elsewhere (e.g., Lambe and Whitman, 1979). The $u(z,t)$ function, that fits equation 7.11 for a loading $\Delta \sigma_v$ in time t, is a Fourier series:

$$u(z,t) = \sum_{m=0}^{\infty} \frac{2 \, \Delta \sigma'_v}{M} \sin \frac{Mz}{H_d} \exp\left(-M^2 T_v\right) \tag{7.12}$$

where: $M = 0.5\pi \, (2m + 1)$, $m = 1, \, 2, \, 3,...$; H_d is the drainage path, corresponding to the maximum path to be followed by water particle A until it reaches the drainage boundary (figure 7.2); T_v is the *time factor*, given by:

$$T_v = \frac{c_v t}{H_d^2} \tag{7.13}$$

Local degree of consolidation

The local degree of consolidation U_z, as a function of depth z and time t, is defined as:

$$U_z = 1 - \frac{\Delta u_t}{\Delta u_{t=0}} \tag{7.14}$$

where: Δu_t is the pore pressure increment at a time t and $\Delta u_{t=0}$ is the initial value corresponding to $t = 0$. Accordingly, U_z is nil at the start of consolidation and equal to 1 or 100% for an infinite time.

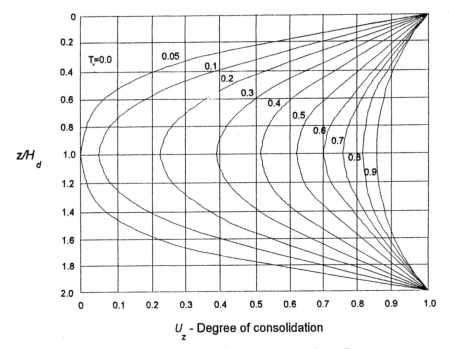

Figure 7.3. Local degree of consolidation U_z as a function of the time factor T_v

Equations 7.14 and 7.13 enable to obtain U_z as a function of relative depth z / H, for several values of the time factor T_v, as shown in figure 7.3. The curves shown are called *isochrones*, corresponding to a unique time (*chronos*, from the Greek meaning time).

Exercise 7.1

A sandy embankment was placed on the soil profile of figure 7.4 and applied a 100 kPa surcharge at the original ground level. Obtain: (i) the drainage path H_d; (ii) the initial pore pressure increment at the middle of the clay layer; (iii) as above, but three years after placement; (iv) as above, but 2 m below ground level. Given: $c_v = 2 \ \text{m}^2 / \text{year}$.

Solution

(i) *Drainage path H_d*. As a sand layer occurs beneath the clay stratum, double drainage can be assumed. The longest path the water particle will follow is from the centre of the clay layer to either the top or bottom boundary. Accordingly, $H_d = H / 2 = 10 \ \text{m} / 2 = 5 \ \text{m}$.

(ii) *The initial value of the pore pressure increment* ($\Delta u_{t=0}$). The breadth of the loading is large enough so that stress increment $\Delta \sigma_v$ is constant with depth. According to Terzaghi's theory: $\Delta u_{t=0} = \Delta \sigma_v$, i.e., $\Delta u_{t=0}$ is considered equal to the surcharge load. Therefore: $\Delta u_{t=0} = 100 \ \text{kPa}$.

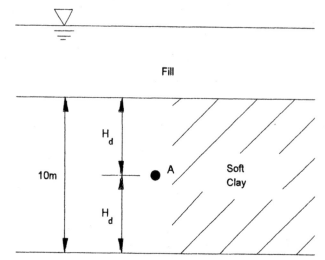

Figure 7.4. Soil profile, exercise 7.1

(iii) Δu_t for t = 3 years in the middle of the clay layer. The corresponding time factor is obtained through equation 7.13:

$$T_v = \frac{2 \ \text{m}^2/\text{year} \times 3 \ \text{years}}{5^2} = 0.24$$

As z/H_d = 5m/5m = 1, figure 7.3 is entered using the isochrone corresponding to T_v= 0.24, interpolating among T_v = 0.2 and 0.3. Therefore, an abscissa value of $U_z \cong 0.33$ is obtained. Employing equation 7.14:

$$\Delta u_t = \Delta u_{t=0}(1 - U_z)$$

$$\therefore \Delta u_t = 100 \ (1 - 0.33) = 67 \ \text{kPa}$$

(iv) *Calculation of Δu_t for t = 3 years and z = 2 m.* Using the same T_v, but with z/H = 2m/5m = 0.4, figure 7.3 is entered again yielding: $U_z \cong 0.60$.

$$\therefore \Delta u_t = 100 \ (1 - 0.60) = 40 \ \text{kPa}$$

Exercise 7.2
Repeat previous exercise. Consider a case of single drainage at the bottom of the clay layer.

Solution

(i) *Drainage path H_d*: In this case the greatest drainage path is the thickness of the clay layer, therefore H_d = 10m.

(ii) The initial pore pressure is the same as in previous exercise: $\Delta u_{t=0}$ = 100 kPa.

(iii) Δu_t for t = 3 years in the middle of the clay layer. First, the time factor is obtained through equation 7.13:

$$\Delta u_t = \Delta u_{t=0}\left(1 - U_z\right)$$

$$T_v = \frac{2 \text{ m}^2/\text{year} \times 3 \text{ years}}{10^2} = 0.06$$

As $z/H_d = 5\text{m}/10\text{m} = 0.5$, $T_v = 0.06$ is entered into the chart in figure 7.3. After interpolation it yields $U_z \cong 0.15$. Using equation 7.14:

$$\therefore \Delta u_t = 100 \ (1 - 0.15) = 85 \text{ kPa}$$

(iv) Calculation of Δu_t for $t = 3$ years, and for $z = 2\text{m}$: Input data for the chart in figure 7.3 are: $z/H_d = 2\text{m}/10\text{m} = 0.2$ and the same time factor as before. Therefore, one gets $U_z \cong 0.55$. Thus:

$$\therefore \Delta u_t = 100 \ (1 - 0.55) = 45 \text{ kPa}$$

Average degree of consolidation

The average degree of consolidation U for the clay layer is calculated by integration of the local value U_z along depth. This means that the value of U corresponds to an area inside an isochrone for a certain T_v value, as shown in figure 7.5a. Therefore:

$$U = \frac{1}{2H_d}\int_0^2 U_z \ dz \tag{7.15}$$

Calculating U for several values of T_v, the relationship $U = f(T_v)$ is worked out and is presented in figure 7.5b and in table 7.1. Curve fitting techniques

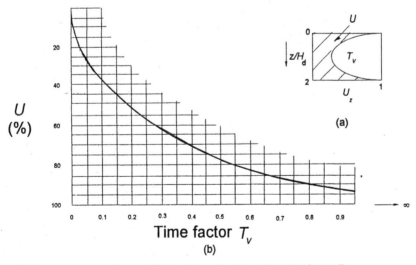

Figure 7.5. Average degree of consolidation U as a function of the time factor T_v

Table 7.1. $U = f(T_v)$ values from Terzaghi's theory for constant initial excess pore pressure distribution with depth

U (%)	T_v
0	0
10	0.0077
20	0.0314
30	0.0707
40	0.126
50	0.196
60	0.286
70	0.403
80	0.567
90	0.848
100	∞

Table 7.2. Approximate relationships $U = f(T_v)$

Equation		Validity
$U = 1.155\, T_v^{0.5}$	(7.16)	$U < 33\%$
$U = 1 - 0.67 \exp\left(0.25 - 3\, T_v\right)$	(7.17)	$U > 33\%$
$U = \sqrt[6]{\dfrac{T_v^3}{T_v^3 + 0.5}}$	(7.18)	$0 < U < 95\%$

(e.g., Atkinson and Bransby, 1978) have led to the approximate useful relationships presented in table 7.2.

The average degree of consolidation can also be defined as a relationship between settlement ρ_t at time t and total settlement at infinite time ρ_∞:

$$U = \frac{\rho_t}{\rho_\infty} \tag{7.19}$$

Exercise 7.3

Compare results of the average degree of consolidation U obtained by Terzaghi's theory and those calculated through the approximate equations of table 7.2 for $T_v = 0.03$.

Solution

The corresponding Terzaghi's theory value (from table 7.1) is:

$$U = 0.20$$

Equation 7.16 yields:

$$U = 1.155 \times 0.03^{0.5} = 0.20$$

The difference between this value and the one in Terzaghi's equation is negligible.

On the other hand, equation 7.18 yields:

$$U = \sqrt[6]{\frac{0.03^3}{0.03^3 + 0.5}} = 0.22$$

Therefore, there is a 10% difference between this value and the previous one given by Terzaghi's solution.

Exercise 7.4

For the profile of figure 7.4, obtain the time for 20% of the settlements to occur. Given: $c_v = 2 \text{ m}^2 / \text{year}$.

Solution

From chart of figure 7.5b, given $U = 20\%$, one obtains $T_{20} \cong 0.03$. Owing to double drainage conditions the drainage path is $H_d = 10 \text{ m}/2 = 5 \text{ m}$. Thus, from equation 7.13:

$$t = \frac{T_v H_d^2}{c_v}$$

$$\therefore t_{20} = \frac{0.03 \times 5^2}{2} \cong 0.4 \text{ years}$$

Exercise 7.5

For the soil profile in figure 7.4, the total settlement caused by a particular load was calculated as 1.2m. Obtain time versus settlement curve. Given: $c_v = 2 \text{ m}^2/\text{year}$.

Solution

Computations are summarized in table 7.3. The first column shows selected U

Table 7.3. Time x settlement data, exercise 7.5

U	ρ_t	T_v	t
(%)	(m)		(years)
20	0.24	0.031	0.4
40	0.48	0.0126	1.6
60	0.72	0.286	3.6
80	0.96	0.567	7.1
100	1.20	∞	∞

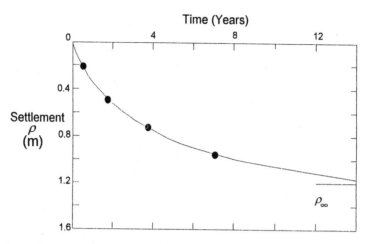

Figure 7.6. Settlement x time curve, exercise 7.5

values. The second, is worked out bearing in mind that, for $U = 100\%$, the total settlement ρ_∞ is 1.2 m. The third column is obtained from table 7.1 or the approximate equations in table 7.2. Finally, in the fourth column, time t is calculated using equation 7.13 with $H_d = 5$ m (double drainage). The time versus settlement curve is presented in figure 7.6.

Solutions for initial pore pressure varying linearly with depth

An important contribution to Terzaghi's theory was the work of Ortenblad, who developed original solutions for boundary conditions not covered in the *Erdbaumechanik*. (Ortenblad A., 1925, *Mathematical theory of the process of consolidations of mud deposits*, ScD thesis, MIT).

In 1922 Alberto Ortenblad was a young Brazilian student at the MIT Department of Civil Engineering. He liked Mathematics and planned to work on structural engineering. A series of facts changed the course of his research. Once, the Dean of the Department called him to his office and asked him to attend lectures of a foreign professor because some students were complaining that it was difficult to understand the mathematical treatment. Alberto was asked to attend Terzaghi's lectures and to report to the Dean. He did so until one day Terzaghi came in very upset and before he began his lecture he said: '*I was told that there is a spy here and I want to know who he is*'. Surprisingly, Alberto Ortenblad stood up and went to see Terzaghi after the class. This was the first of many fruitful discussions and a lifelong friendship. Alberto decided to work with Terzaghi on the theory of consolidation. At that time, only the simplest cases of constant pore pressures increments with depth had already been solved. He worked on solutions for different distributions of pore pressures with depth. His research caused to be the first doctorate degree in engineering to be awarded at MIT. Engineers, up to that time, were not supposed to produce anything original enough to deserve a doctor's degree. At his thesis' defense , more than 20 MIT professors formed the committee and asked him with all sorts of questions. Terzaghi came in support of his Brazilian student. At the end, his *ScD* degree was awarded.

These conditions contemplate cases in which the initial pore pressure distribution is not constant with depth. In fact, field measurements of pore pressure

Table 7.4. Exact solutions of Terzaghi's consolidation equation for the initial pore pressure $\Delta u_{t=0}$ distribution varying linearly with depth

U (%)	T_v	T_v
0	0	0
10	0.04	0.003
20	0.10	0.009
30	0.15	0.024
40	0.22	0.048
50	0.29	0.092
60	0.38	0.160
70	0.50	0.271
80	0.66	0.44
90	0.94	0.720
100	∞	∞

show that its distribution is seldom constant with depth. Ortenblad's works were later included in many text-books (e.g., Taylor, 1948, Leonards, 1962) and two cases are reproduced in table 7.4. Case 1, corresponds to the situation in which the initial pore pressure $\Delta u_{t=0}$ is nil on the surface, and case 2, nil at the bottom of the layer.

A few additional solutions are available for different initial pore pressure distributions and can be found elsewhere (e.g., Ortigao and Almeida, 1988).

Settlement types

Deformation during oedometer compression can be caused by different phenomena. Consider the results of figure 7.7 showing one loading stage in an oedometer test, in which settlements are plotted against logarithm of the elapsed time. It is possible to distinguish three types of settlement: *initial*, *primary* and *secondary*.

Initial or *immediate settlements* occur simultaneously with the application of loading. This is due to compression of gases, if the soil is not fully saturated, or due to shear deformation that affects vertical displacements. Close to the toe of a

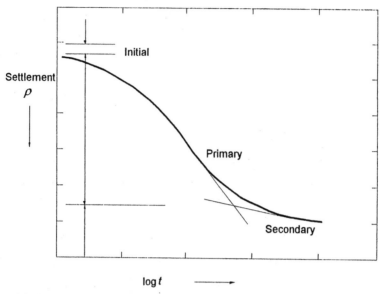

Figure 7.7. Initial, primary and secondary settlements

loaded area, shear deformation is important and may affect vertical displacements. This is also the case of a small loaded area compared with the thickness of the soft foundation.

Primary settlements are consolidation settlements, considered in Terzaghi's theory, due to water expulsion from void spaces in soil.

Secondary settlements, also known as *drained creep*, consists of slow deformation of soil skeleton with time, and may occur at constant effective stresses. They correspond to the final portion, i.e., the final straight line of the settlement versus log time curve (figure 7.7), in which deformations are proportional to the logarithmic of time.

The above classification of settlement types has only didactic purposes, to facilitate understanding and modeling of the phenomena. However, the three types may occur simultaneously.

Primary settlements can be treated the same way as any problem of stress and deformation in the continuum. Elasticity may be applied at the beginning of the loading, when the safety factor is sufficiently high, and proportionality of stress and strain can be assumed. Therefore, soil behaviour is represented by elastic parameters E and v. Elastic settlements can be evaluated by means of charts, like in the Poulos and Davis' (1974) book.

In most of cases, secondary settlements have little influence on the behaviour of a structure, because their magnitude is considerably smaller than the other settlement types.

Settlement measurements in foundation of structures the classification of soils types according to secondary settlement.

Sands: In sands, creep is of little importance, since buildings on sands present settlements that cease immediately after loading.

Clays: In clays, the secondary settlement can be significant. Designers usually allow 5 to 10% of the estimated total settlement for secondary settlements. As secondary settlement measurements in the laboratory tend to be proportional to the logarithm of time, its magnitude decreases for each log cycle, but theoretically does not cease. Experience in the observation of old buildings on shallow foundations on clay in Rio de Janeiro (Nunes, 1971) indicated settlement rates as low as 1 μm per day.

Peats: Peats are deposits of fibrous organic matter like roots, that frequently present exceptionally high water contents, sometimes reaching 1000%. The void ratio can reach values in the order of 20 (e.g., Casagrande, 1966 and Perrin, 1973). Secondary settlements can be very important in these soils. The compressibility and initial permeability of peat are very high due to the high percentage of void spaces. If loading is applied, the resultant excess pore pressure dissipates very rapidly, in a matter of minutes. It is followed by the secondary settlement phase, during which most of the settlements occur and may take many years.

A practical implication is that any soil stabilization process that accelerates drainage (in order to accelerate settlements) may not work in peats.

Determination of the coefficient of consolidation c_v

The accuracy of settlement rate predictions rely on the coefficient of consolidation c_v, but its correct value is difficult to be evaluated. There are laboratory and in situ testing methods for obtaining c_v, that will be presented and discussed below.

Determination of c_v from oedometer tests

Results of a loading stage in an oedometer test allow the determination of c_v using the traditional methods of Casagrande, or log t, and Taylor, or $\sqrt{t}$. Both methods were developed from curve fitting techniques to adjust laboratory results to the theoretical time versus settlement curve.

Casagrande or log t method

Figure 7.8 presents the results of a 160 kPa loading stage on a Rio de Janeiro clay specimen. The abscissae correspond to the elapsed time in minutes in a log scale. The ordinates, to the vertical displacement. Time readings were taken in a geometric progression in order to match the log scale. Therefore, the elapsed times in minutes were: 0.1; 0.25; 0.5; 1; 2; 4; 8; 15; 30; 60; 120; 240; 480 and 1440 (24 hours). The ordinates correspond to vertical displacement in millimetres.

The steps to obtain c_v are:

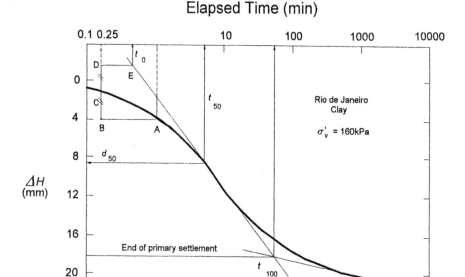

Figure 7.8. Determination of c_v by the log t method

1. Plot a straight line tangent to the final part of the curve, corresponding to the secondary settlement.
2. Plot another straight line tangent to the central portion of the test curve, through its inflection point.
3. The interception between the two straight lines corresponds to the *end of primary* settlement, at a time t_{100}.

The following additional steps are employed to fit a parabola to the initial portion of the test curve, determining its asymptote:

4. Determine point A on the test curve, corresponding to the abscissa of 1 minute.
5. From A, plot a horizontal line which has point B at the abscissa 0.25 minutes.
6. Plot C at the same 0.25 minute abscissa, but on the test curve.
7. Point D, also having an a 0.25 minutes abscissa, is determined considering that segment BC is equal to CD.
8. The horizontal line the contains D is the asymptote to the parabolic curve fitted to the initial part of the test curve. Its interception E yields t_0. The centre of the segment between t_0 and t_{100}, has coordinates t_{50} and ΔH_{50}, respectively the time and the displacement at 50% consolidation. Then:

$$t_{50} = 5.6 \text{ min} = \frac{5.6}{60 \times 24 \times 365} = 1.07 \times 10^{-5} \text{ years}$$

$$h_{50} = \frac{1}{2}\left(H_0 - \Delta H_{50}\right)$$

where: H_0 = initial height of the specimen, equal to 14 mm; ΔH_{50} = displacement obtained at the test curve corresponding to t_{50}, equal to 0.88 mm; h_{50} = drainage path at 50% consolidation;

$$h_{50} = \frac{1}{2}\left(14 - 0.88\right) = 6.6 \text{ mm} = 0.066 \text{ m}$$

9. Finally, applying equation 7.13:

$$c_v = \frac{T_{50}h_{50}^2}{t_{50}} \tag{7.20}$$

where: $T_{50} = 0.196$ (obtained from table 7.1 for $U = 50\%$), then:

$$c_v = \frac{0.196 \times 0.0066^2}{\left(1.07 \times 10^{-5}\right)} = 0.8 \text{ m}^2/\text{year}$$

Taylor's or √t method

Data from each loading stage are plotted as shown in figure 7.9, in which the abscissae are the square root of the elapsed time, $\sqrt{t}$, and the ordinates, the vertical

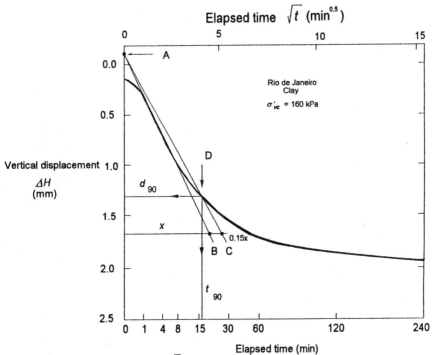

Figure 7.9. Determination of c_v by the $\sqrt{t}$ method

displacements. The steps to obtain c_v are:

1. The typical test curve plots with an initial curved portion followed by a straight line, from which a straight tangent line, fitted between data points, is drawn.
2. Extend the tangent line that intercepts the ordinate axis at *A*.
3. Choose point *B* anywhere along the tangent line. This point corresponds to a time *t*.
4. Point *C* is to the right of point B, corresponding to a $0.15t$ time
5. Draw line *AC*, which crosses the test curve at *D*.
6. The coordinates of point *D* are: t_{90} and ΔH_{90}, respectively: the time and settlement corresponding to $U = 90\%$.
7. The following value is obtained in the test curve: $t_{90} = 16\,\text{min}$, therefore, $t_{90} = 3.04 \times 10^{-5}$ years.
8. Since the given initial height of the specimen is 14 mm and the displacement ΔH_{90} from the test curve is 1.3 mm, the drainage height is:

$$\therefore h_{90} = \frac{1}{2}\,(14 - 1.3) = 6.3\,\text{mm} = 0.063\,\text{m}$$

c_v is then calculated by:

$$c_v = \frac{T_{90} h_{90}^2}{t_{90}} \tag{7.21}$$

where: $T_{90} = 0.848$, from table 7.1.

$$\therefore c_v = \frac{0.848 \times 0.0063^2}{3.04 \times 10^{-5}} = 1.1\,\text{m}^2\big/\text{year}$$

Discussion of methods

The Casagrande method can be easily applied to clays in the normally consolidated stress range, as shown in figure 7.8. However, for the overconsolidated stress range it may be difficult to use this method, as illustrated in figure 7.10. It shows the whole set of volumetric strain x log *t* of curves for one test with several loading stages. The test began with small load increments until it reached the overconsolidation stress of 25 kPa. Then, load increments were doubled. A completely different behaviour is presented by these curves before and after the overconsolidation stress. In the early stages of loading the curves in figure 7.10 are very flat and it does not seem possible to determine the end of primary consolidation. Therefore, it is difficult to work out c_v using Casagrande's method. This problem does not seem to occur with $\sqrt{t}$ method, which is the writer's preferred method.

The methods of Casagrande and Taylor never give the same answer, except by coincidence. In this example the values of c_v were, respectively, 0.8 and 1.1 m²/year, relatively close. This is typical for soft clays, in which differences

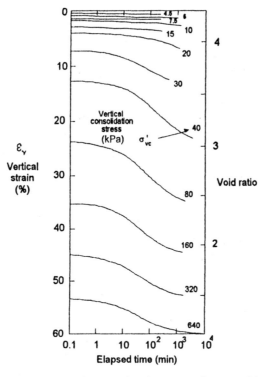

Figure 7.10. Comparison between settlement x log time curves for several loading stages in a oedometer test on soft Rio de Janeiro clay

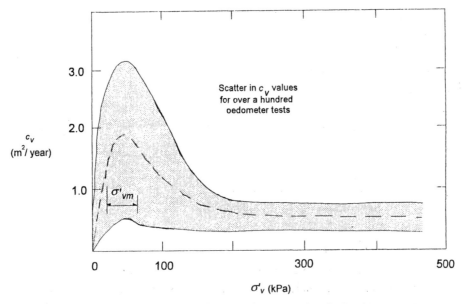

Figure 7.11. Range of c_v values for Rio de Janeiro clay from over a hundred oedometer tests

in c_v values between these methods can vary in the 50 - 150% range (Ladd, 1973).

Different sources of variability in c_v values are considered. One due to the calculation method, another due to sample disturbance and a third to natural soil variability. As an example, c_v in over a 100 tests on Rio de Janeiro clay is shown in figure 7.11 (Ortigao and Almeida, 1988). In the overconsolidation range (σ'_v less than 100 kPa) c_v varies from 1 to 3.5 m²/year. In the normally consolidated range ($\sigma'_v > 100$ kPa) the scatter narrows to 0.5 ± 0.3 m²/year.

These data demonstrate the difficulty of selecting one particular value of c_v from oedometer tests for the application of Terzaghi's theory.

Correlation between c_v and physical properties

An useful universal empirical correlation between c_v and simple index properties was obtained by Carrier (1985).

$$c_v = \frac{28.67}{PI} \frac{\left(1.192 + A_c^{-1}\right)^{6.993} \left(4.135 LI + 1\right)^{4.29}}{\left(2.03 LI + 1.192 + A_c^{-1}\right)^{7.993}} \left(m^2/year\right) \qquad (7.22)$$

where: PI is the plasticity index (%), A_c is the activity and LI is the liquidity index.

Equation 7.22 was obtained for remoulded clays, i.e., completely disturbed, without any structural effect. Remoulding provokes a reduction in c_v, therefore this equation yields a lower bound value in relation to intact clays.

Figure 7.12 presents a chart for graphical solution of equation 7.22.

Exercise 7.6

Evaluate c_v for Rio de Janeiro clay. Given $w = 150\%$, $PI = 80\%$, $PL = 40\%$ and the percentage of clay is 55%.

Solution

The following values are obtained: $A_c = 1.45$ and $LI = 1.38$. Applying equation 7.22:

$$c_v = \frac{28.67}{80} \frac{\left(1.192 + 1.45^{-1}\right)^{6.993} \left(4.135 \times 1.38 + 1\right)^{4.29}}{\left(2.03 \times 1.38 + 1.192 + 1.45^{-1}\right)^{7.993}}$$

$$\therefore c_v \cong 0.5 \ m^2/year$$

This value lies within the normally consolidated range presented in figure 7.11, in which the clay has already lost its structural effect. The chart in figure 7.12 can also be used for this example. Compare the two solutions.

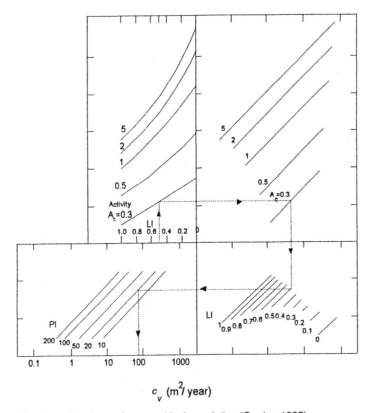

Figure 7.12. Chart for estimating c_v from empirical correlation (Carrier, 1985)

c_v from in situ tests

Difficulties associated with selection of an appropriate value of c_v from laboratory tests have led engineers to seek alternatives, like measurements in situ. This has the advantage of avoiding disturbance during sampling, transportation, storage and preparation, but the benefits of laboratory controlled conditions are lost.

In situ test methods commonly used to obtain c_v are:

(i) piezocone, (ii) settlement analyses through Asaoka's method and (iii) the combined method of measurements of in situ permeability and laboratory compressibility.

Piezocone

Piezocone tests are cone penetration tests (CPTU) in which an instrumented steel cone (figure 7.13) is statically pushed into the ground. The cone has a standard cross sectional area of 10 cm^2, corresponding to 36.6 mm in diameter, and apex angle of 60°, where simultaneous measurements of tip resistance or cone bearing q_c, in MPa or kPa, lateral friction f_s, in kPa, and pore pressure u, in kPa.

The rate of penetration of the cone in soils is standardized as 2 cm/s and

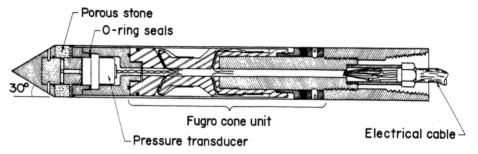

Figure 7.13. Piezocone (Campanella et al., 1983)

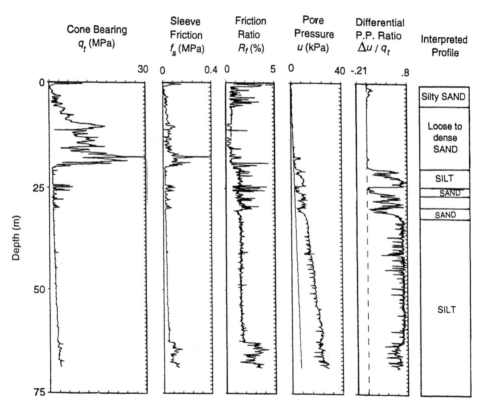

Figure 7.14. Typical CPT results in sand overlying silty clay deposit in Richmond, BC, Canada: sleeve friction f_s, tip resistance q_c and the friction ratio R_f

measurements are practically continuously and automatically recorded along depth. This feature enables a high sensitivity for the detection of variation of strength of a stratified layer, even if a seam is a few centimetres thick. This makes the piezocone the most advanced test for statigraphy, i.e., the description of a sequence of soil layers. This advantage matches a long standing problem in the study of consolidation problems: the identification of thin sand seams in a clay deposit. As a matter of fact, thin sand seams can remain undetected by conventional site investigation techniques, such as sampling though boreholes.

Examples of CPTU logs are presented in figures 7.14 and 7.15, where q_c, f_s, u and the friction ratio $R_f = f_s/q_c$, are plotted along depth. Figure 7.14 presents a profile of dense sand overlying soft silty clay from Richmond, British Columbia, Canada, obtained by the writer with the UBC piezocone equipment. The results show very clearly the sharp transition from sand to clay at approximately 20 m depth. Then, at 25 m depth, thin layers of sand are detected as q_c increases and pore pressure u decreases. Similar results are shown in figure 7.15

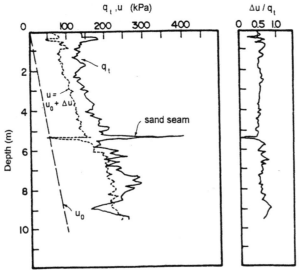

Figure 7.15. Detection of a sand seam in soft Rio de Janeiro clay through CPTU (Rocha et al., 1985)

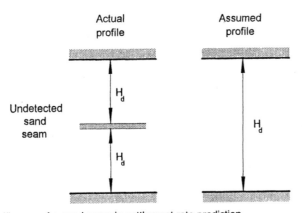

Figure 7.16. Significance of a sand seam in settlement rate prediction

for the soft Rio de Janeiro clay, in which a sand seam was clearly detected by the piezocone at 5.5 m depth.

The presence of undetected sand seams would make any consolidation prediction completely wrong, because the drainage path length would be different of the one used in the calculations, in addition, the wrong value is squared in equation 7.13, producing an enormous error. This is shown in figure 7.16.

As an example, the writer has been involved in a breakwater project in which the designer used c_v for the clay foundation of 20 m^2/year, about three times the laboratory values, and justified his choice supposing the presence of sand seams. A CPTU programme was, then, carried out and no sand lenses were detected. This led to new settlement rates prediction with reduced c_v values, which were later confirmed by field observations during construction (Ortigao and Sayao, 1994).

CPTU testing can be carried out from a truck (figure 7.17) for greater productivity and quick mobilization, since the truck self weight, up to 200 kN, is used as reaction against penetration resistance.

The determination of the coefficient of consolidation is carried out at certain depths by interrupting cone penetration and observing pore pressure dissipation around the cone tip, as the example in figure 7.18. Field results are then compared to a theoretical solution allowing the determination of the coefficient of consolidation. A detailed discussion of the procedures that can be employed are out of the scope of this book, but can be found elsewhere (e.g., Robertson and Campanella, 1989, Meigh, 1987, and Ortigao and Almeida, 1988).

The dissipation test can be analysed in the following simple way:

$$c_h = \frac{T r^2 I_r^{0.5}}{t} \tag{7.23}$$

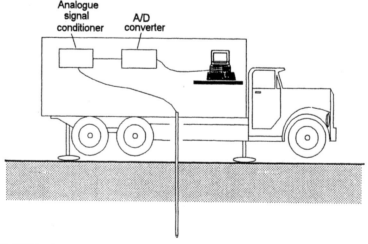

Figure 7.17. CPT truck

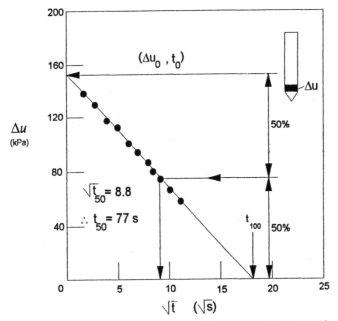

Figure 7.18. Example of pore pressure dissipation around the cone and the method for obtaining t_{50}

where: c_h is the coefficient of consolidation in the horizontal direction, r is the standard piezocone radius, 18.3 mm, I_r is the rigidity index given by $I_r = G/c_u$, where G is the soil shear modulus and c_u its undrained shear strength, which will be studied later in chapter 12. For most soft clays a value of I_r between 50 to 100 can be employed. T is the time factor from Houlsby and Teh's (1988) solution of dissipation around the cone. They quote a value $T = 0.245$ for dissipation measured just behind the cone, as shown in figure 7.13. Then, using equation 7.23, the value of c_h can be obtained as follows:

$$c_h = \frac{\dfrac{0.245 \times 0.0183^2 \times 100^{0.5}}{77}}{3600 \times 24 \times 365} = 336 \text{ m}^2 \text{ / year}$$

The value of c_h from dissipation tests refers to the recompression range, therefore must be reduced if one wants to use it in flow problems in the normally consolidated range, as discussed further by Robertson and Campanella (1989).

Asaoka's method

Asaoka (1978) devised a brilliant and very simple method for obtaining the coefficient of consolidation from analysis of field measurements of settlements.

In major projects it may be worthwhile to carry out field scale experiments to test different types of engineering solutions and to obtain soil parameters from the observed behaviour of the test structure. This can be the case of a road built on soft foundation, a dam, or a harbour. A few cases in which the writer has been involved (Ortigao et al., 1983 and Almeida et al., 1988) yielded significant data for the design of roads in Rio de Janeiro and an embankment dam (Coutinho and Ortigao, 1990).

For these cases, Asaoka's method is an invaluable tool to analyse settlements. Its theoretical basis will not be presented here, only the steps for its application:
1. Settlement versus time observations are plotted as shown in figure 7.19a.
2. The time scale is divided into constant Δt intervals. This value is usually taken between 15 to 100 days, depending on the data available. For each Δt interval, settlement values ρ_1, ρ_2, ρ_3, .. ρ_n are determined in the plot, corresponding to times t_1, t_2 t_3, t_n.
3. Settlements ρ_i, at times t_i, are then plotted versus settlements ρ_{i-1}, corresponding to times t_{i-1}, as shown in figure 7.19b. A 45° line is drawn.
4. A line is fitted through the experimental data until it intercepts the 45° line. This yields the total settlement ρ_∞. Angle β_1 enables the determination of the coefficient of consolidation c_v through the following equation, valid for double drainage.

$$c_v = -\frac{5\,H_d^2\,\ln\beta_i}{12\,\Delta t}$$

(7.24)

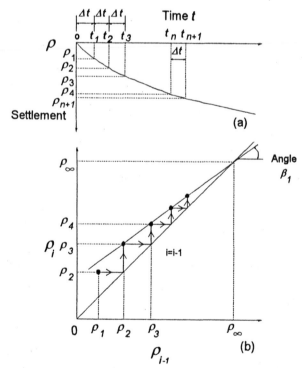

Figure 7.19. Asaoka's method for analysis of settlements: (a) settlement versus time data points for a constant time interval Δt; (b) determination of the total (final) settlement ρ_∞ and the parameter β_1

Exercise 7.7

Use Asaoka's method to analyse the observed settlements at Juturnaiba dam, RJ (figure 7.20), on soft foundation. Data are plotted in figure 7.21 and tabulated in table 7.4.

Solution

Time interval Δt was selected as 25 days.

A third column was added to table 7.5 with ρ_{i-1} data, to facilitate plotting. Then, figure 7.22 was plotted and a linear regression on the data yielded:

$$\rho_i = 153.5 + 0.83\,\rho_{i-1}$$

The interception of this line with the 45° line yields the total settlement ρ_∞ of 908 mm.

The slope of the regression line is $\tan\beta_1 = 0.83$, then: β_1 is 0.69 radians.

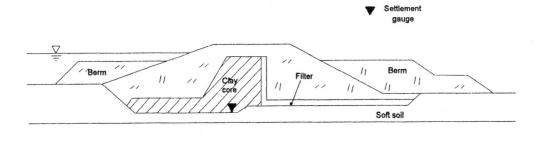

Figure 7.20. Cross section of the Jururnaíba dam, Brazil

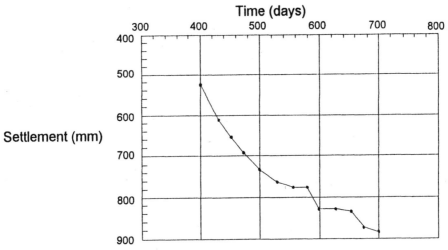

Figure 7.21. Settlements-time measurements at Juturnaíba dam

Table 7.5. Settlement versus time data for the application of Asaoka's method to Juturnaiba dam

Time	ρ_i	ρ_{i-1}
(days)	(mm)	(mm)
400	530	
425	610	530
450	650	610
475	690	650
500	720	690
525	750	720
550	770	750
575	770	770
600	820	770
625	820	820
650	830	820
675	870	830
700	880	870

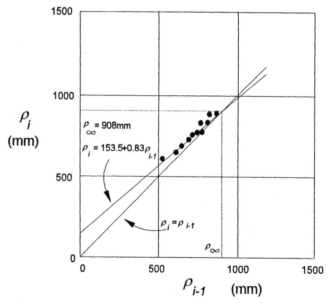

Figure 7.22. Asaoka plot for the analysis of settlements of the Juturnaiba dam

The coefficient of consolidation is obtained considering double drainage and a thickness of the soft layer of 4.5 m (figure 7.20):

$$c_v = -\frac{5\left(4.5/2\right)^2 \ln 0.69}{\dfrac{12 \times 25}{365}} = 11 \ \mathrm{m^2/year}$$

Combined method

The *combined method* consists of utilizing equation 7.9, which gives c_v from permeability k and soil modulus m_v (or the Janbu's modulus M). The name *combined* refers to permeability values obtained from in situ tests and soil modulus from laboratory tests or other in situ tests.

In situ permeability tests can be conducted in several ways (e.g., Cedergren, 1977). In soft soils the permeability k can be evaluated through piezometers, such as the Casagrande piezometer described in chapter 3, by carrying out a variable head permeability test. It consists of raising the water level inside the access pipe (figure 7.23) and observing its change with time. Further discussion about in situ permeability is out of the scope of this text, but can be found elsewhere (e.g., Daniel 1989, Tavenas et al.,1986, Leroueil et al., 1985, and in the classical work of Hvorslev, 1951).

The permeability is calculated from:

$$k = \frac{a}{F\left(t_2 - t_1\right)} \ln\frac{h_1}{h_2} \tag{7.25}$$

where: h_1 and h_2 are water head observations in the piezometer at times t_1 and t_2, a is the cross section area of the piezometer pipe and F is the shape factor of the instrument, given by:

$$F = \frac{2\,\pi\,L}{\ln\left(\frac{L}{D} + \sqrt{1 + \left(L/D\right)^2}\right)} \tag{7.26}$$

where terms are defined in figure 7.23.

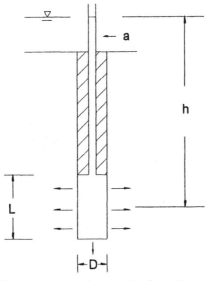

Figure 7.23. In situ permeability tests through Casagrande piezometer

In situ permeability tests have the advantage of testing a bigger mass of soil and can include the effect of non-homogeneity such as sand lenses and fissures, impossible to account for in laboratory tests. If the test is carried out through a piezometer, like the one shown in figure 7.23, the flow of water is essentially radial, and so is the permeability value. This fact is of little concern for homogeneous deposits.

Comparison between in situ and laboratory methods

A comparison between values of c_v from laboratory and in situ testing methods is given in figures 7.24 to 7.26 for three deposits: the Rio de Janeiro clay, a silty clay deposit near Vancouver and an offshore deposit at the northern Brazilian coast. The coefficient of consolidation obtained from Asaoka's method is the reference datum because it has been obtained from backanalysis of field data and allows to reproduce measured behaviour for a particular deposit. The figures also show that there is not a unique value of c_v, but it depends on the testing method. Therefore, the best way to obtain c_v is to use a particular testing method and to correct the results according to previous calibration data.

Proposed exercises

7.1. Define: (a) coefficient of consolidation; (b) time factor; (c) average and local degree of consolidation.

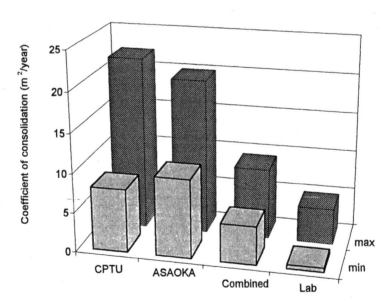

Figure 7.24. Comparison of coefficient of consolidation of the Rio de Janeiro clay from various methods (data from Almeida et al., 1989, and Danziger, 1990)

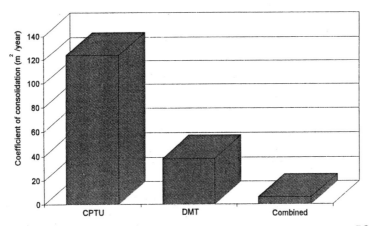

Figure 7.25. Comparison of coefficient of consolidation of a silty clay from Vancouver, BC, Canada from various methods (data from UBC test records)

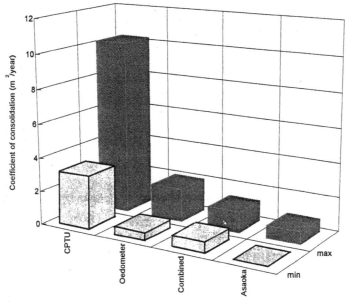

Figure 7.26. Comparison of coefficient of consolidation of an offshore deposit, northeastern Brazil (Ortigao and Sayao, 1994)

7.2. Explain why sands present immediate consolidation while in clays the process may be very slow.

7.3. What is the Terzaghi's piston-spring-water system analogy?

7.4. List and briefly discuss Terzaghi's hypotheses on consolidation theory.

7.5. Data in the following table refer to a consolidation test in which the vertical load was 40 kPa. Obtain: (a) c_v through $\log t$ and $\sqrt{t}$ methods; (b) permeability k from equation 7.9, in which m_v can be obtained from data on Rio de Janeiro clay in figures 6.8 and 6.9. Given: the initial specimen height was 25 mm.

Elapsed time (min)	Readings (mm)
0	4.041
0.1	3.927
0.25	3.879
0.5	3.830
1	3.757
2	3.650
4	3.495
8	3.282
15	3.035
30	2.766
60	2.550
120	2.423
240	2.276
505	2.184
1485	2.040

7.6. A 4.5 m high sand embankment with unit weight 19 kN/m³ will be placed on Rio de Janeiro clay (compressibility from chapter 6, c_v from figure 7.11). Obtain: (a) total settlement; (b) time-settlement curve; (c) estimate pore pressure response for a piezometer installed at 2 m depth, and another at 5 m depth.

7.7. The following table presents time settlement data from measurements at Juturnaiba dam (figure 7.21). Apply Asaoka's method to predict total settlement and the value of c_v.

Time (days)	Settlement (mm)
400	700
425	750
450	780
475	800
500	840
525	860
550	870
575	880
600	920
625	940
650	950
675	995
700	1100

7.8. A Casagrande piezometer having 37 mm diameter and a 600 mm high porous filter was pushed into clay until it reached a 25 m depth. After leaving it to rest for a month, a variable head in situ permeability test was carried out. The water level tidal fluctuation in the instrument was observed for the first 47 hours and then the water column was raised by adding 1 m of water. Dissipation was then observed with time. Data are summarized in the following table (water column h in the piezometer versus elapsed time t). The following results are required: the permeability of the clay and an estimate of the coefficient of consolidation. Use m_v from Rio de Janeiro clay (chapter 6) for the in situ stress range. *Suggestion*: use a spreadsheet program to plot h versus time and evaluate equilibrium head h before the test. The test starts at a time $t = 47$ hours. Then, using equation 7.25 calculate the variation of permeability versus time. Select the appropriate value of k and then, compute c_v.

t (hours)	h (m)	t (hours)	h (m)	t (hours)	h (m)
0	21.60	63	22.07	115	21.58
5	21.58	64	22.02	116	21.58
6	21.60	73	21.92	117	21.58
14	21.57	74	21.90	128	21.56
18	21.57	75	21.85	129	21.57
20	21.61	79	21.79	130	21.57
27	21.58	88	21.75	131	21.58
29	21.60	89	21.73	139	21.56
30	21.64	90	21.72	140	21.55
38	21.62	91	21.70	141	21.54
39	21.58	92	21.68	142	21.56
41	21.59	100	21.67	143	21.57
42	21.60	101	21.66	150	21.57
44	21.62	102	21.64	151	21.56
47	22.64	103	21.64	152	21.55
47	22.64	104	21.64	153	21.54
47	22.61	105	21.65	154	21.54
48	22.60	112	21.65	155	21.55
48	22.57	113	21.63	156	21.56
49	22.53	114	21.59	162	21.57

Chapter 8

Laboratory test types

Introduction

Laboratory tests are the basic tools for the study of stress-strain strength properties of soils. The oedometer test was reviewed in chapter 6. The present chapter focuses on additional types of tests such as the triaxial and the direct shear ones. The following aspects will be analysed: stress paths, drainage conditions, equipment and advantages and disadvantages of each type.

Test types

Figures 8.1 and 8.2 summarize the main soil test types, specimen deformation and stress paths, which will be discussed below:

Isotropic compression test

The sample in figure 8.1a is subjected to an all around pressure σ_c. The stress path during isotropic compression coincides with the hydrostatic axis. These loading conditions seldom apply to field situations and this test has little application in soil mechanics. There are a few exceptions, however. As an example, in the study of grain crushing in granular soils (chapter 6), a specimen is compressed under very high pressures of the order of tens MPa. In this case, an isotropic stress cell may be preferred because it is easier to design for very high pressures.

Oedometer compression test

In oedometric compression (figure 8.1b) the specimen is confined preventing lateral strains: $\varepsilon_2 = \varepsilon_3 = 0$. This simulates deformation conditions encountered in situ during the formation of sedimentary soils. The effective stress path follows a K_0 relationship, as studied in chapter 6. The oedometer test is used for the study of stress and strain properties before failure.

164

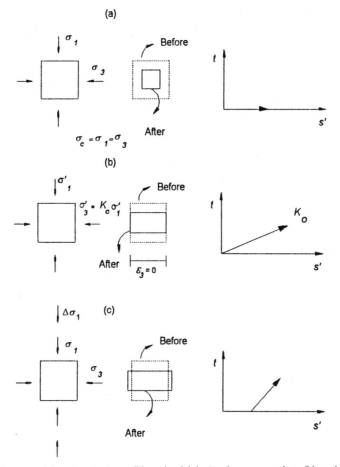

Figure 8.1. Common laboratory test conditions in: (a) isotropic compression; (b) oedometer compression; (c) triaxial compression.

Triaxial compression test

The triaxial compression test (figure 8.1c) is carried out on a cylindrical specimen in which axial and radial stresses can vary. The test name *triaxial* is inappropriate because the stress system is axisymmetrical, and not truly triaxial.

The test is generally conducted in two phases. First, an isotropic confining stress σ_c is applied. This is followed by a shear phase in which axial stress increases through the application of a deviator stress $\Delta\sigma_1 = \sigma_1 - \sigma_3$, until failure occurs. In the shear phase, σ_c is usually constant. The stress path has two portions, the first is horizontal, corresponding to the application of the confined stress. The second is a 45° inclined path, corresponding to the increase of the deviator stress.

This test is used in the study of soil strength and the deformation behaviour before failure. It is a versatile type of test, allowing controlled stress paths for simulating certain loading conditions, as will be studied later in chapter 13.

Direct shear test

The direct shear (figure 8.2a) was the first soil test. It was devised by Coulomb in 1776 (*Essai sur une application des regles de maximis et minimis à quelques problèmes de statique relatifs à l'architectutre*, Mémoires Academie Royale, pres. divers Savents, 7, Paris, 38 p.) for the study of shear strength. The specimen is placed in a split box. A normal force N is applied, followed by a tangential force T. One part of the box will slide in relation to the other, until soil specimen fails. The normal and shear stresses on the failure plane are: $\sigma = N / a$ and $\tau = T / a$, where a is the cross sectional area of the soil specimen.

In the first phase of the test, during the application of the normal stress, deformation conditions in the split box are similar to those in the oedometer test. A K_0 stress path is followed by the soil specimen.

Once shearing starts, stresses and strains in the specimen become nonuniform. It is then impossible to determine linear strains or distortions from measurements taken outside the specimen. The only possibility for assessing strains is the use of internal gauges in a very large specimen, as in the research carried out by Palmeira (1987) on 1 m x 1 m x 1 m specimens. Since this is not

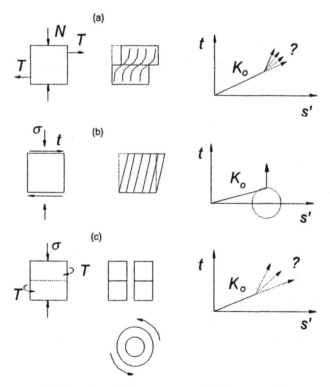

Figure 8.2. Stresses and deformations in laboratory tests: (a) direct shear; (b) simple shear; (c) ring shear

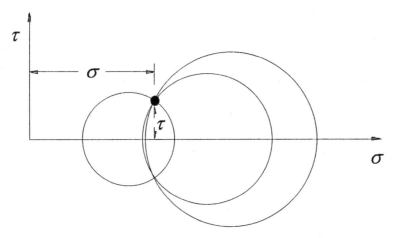

Figure 8.3. Possible Mohr's circles for a direct shear test results, in which the stresses at only one plane are known

possible in conventional tests, no information on the state of strain or stress is obtained. It is impossible to obtain deformation parameters such as Young's modulus and Poisson's ratio. The only information from direct shear tests are normal and shear stresses on the failure plane.

The results of a direct shear test plot as a single point on a Mohr's diagram (figure 8.3), allowing several Mohr's circles to be obtained. Therefore, the state of stress is indeterminate.

Simple shear
The simple shear test (figure 8.2b) was an improvement in relation to the direct shear because it submits the specimen to a uniform state of stress and strain. The test is conducted by applying a normal stress σ in oedometric conditions, allowing consolidation along the K_0 line. Then, a shear stress is applied and distortions take place until the specimen fails.

The simple shear was developed at the Norwegian Geotechnical Institute NGI (Bjerrum and Landva, 1966) and later at Cambridge University (Roscoe, 1970). Test equipment is much more complex than the direct shear, and the great advantage of this test, simplicity, is lost.

Ring shear
The ring shear test (figure 8.2c) allows a normal stress to be applied under oedometric conditions, followed by a controlled torsion. Failure occurs in a predetermined plane in the ring containing the specimen. The test is used for the study of shear strength under very large deformation, of the order of a metre. It is possible to apply many rotations between the upper and bottom parts of the specimen and to observe the degradation of strength. Ultimately a residual shear strength condition may be obtained. This represents the shear strength occurring,

for instance, in the slip plane of a landslide. This point will be discussed further in chapter 13.

A relatively simple equipment that allows the test to be used in practical applications was designed by Bromhead (1979). Test techniques are discussed by Bromhead and Curtis (1983) and Bromhead (1986).

Other types of tests

There are many complex soil test apparatuses used exclusively in research. The *true triaxial* attempts to vary σ_1, σ_2 and σ_3 independently on cubical specimen. The *plane strain* triaxial imposes $\varepsilon_2 = 0$ condition, and the *hollow cylinder* apparatus (e.g., Sayao and Vaid, 1988) enables stress rotation without changing stress magnitude. These tests will not be discussed here.

Exploring the direct shear test

The direct shear test is very common in practice, therefore will be discussed in more detail.

Equipment

The cell is shown in figure 8.4. It comprises a split steel box containing the specimen. Drainage is allowed at the top and bottom by means of porous stones. Normal force is distributed on the top of the soil specimen by means of a rigid steel plate. The specimen can be maintained under water to avoid loss of water content and of saturation.

The tangential force is applied at the bottom part of the split box leading to displacements in relation to the upper part, which is not allowed to move. Linear ball bearings support the bottom part and eliminate friction.

The lateral force is measured by means of an electrical or mechanical trans-

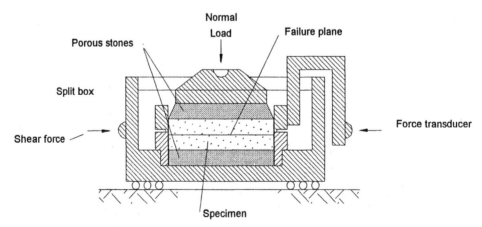

Figure 8.4. Details of the shear box

ducer. Displacement gauges are employed to observe vertical and horizontal displacements of the specimen.

Details of specimen preparation and testing are out of the scope of this text, but are discussed by laboratory testing manuals, such as Vickers's (1978), or the comprehensive treaty by Head (1980).

Specimen drainage

Drainage should be allowed throughout the test, since there is no way of sealing the specimen. In fact, even if porous stones were replaced by steel plates, it would not be possible to control drainage in the direct shear. Once the shear phase starts and the upper part of the specimen moves in relation to the bottom, a lateral gap opens. Water can flow through this gap. Drainage control becomes impossible. The only solution is, therefore, to allow full drainage throughout the test, and keep excess pore pressures equal to zero.

Drained condition implies full dissipation during shear. In sands, due to their high permeability, this is immediate, and shearing may be conducted quickly in ten minutes or less. In clayey soils, full drainage may require long testing time to allow for pore pressure dissipation.

Some commercial laboratories still attempt to conduct undrained direct shear tests in clayey soils. Shearing is carried out very quickly in a few minutes, and the test is supposed to be undrained. This is absolutely unacceptable and leads to totally erroneous results.

Failure plane

The failure plane in the direct shear is imposed on a horizontal surface. This can be a disadvantage when dealing with homogeneous soils, in which a weak plane direction cannot be detected a priori, and during specimen trimming the wrong direction is chosen, yielding to unconservative results.

On the other hand, there are cases in which it is interesting to evaluate the shear strength as a function of the direction of the failure plane. As an example, *structured residual soils* are materials that still keep discontinuities such as joints and fissures from the bedrock. This is common in residual soils from metamorphic rocks presenting shear and bedding planes, such as slates, hematite and quartzites. Failure in these soils will always occur along distinct weak bedding or shear planes.

A typical example is shown if figure 8.5. Two specimens *A* and *B* were trimmed from a block sample of a structured soil. Specimen *A* is parallel and *B* perpendicular to the bedding planes. The direct shear becomes a useful tool to analyse strength variation with inclination in relation to weak planes.

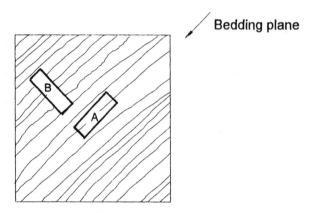

Figure 8.5. Soil sample presenting bedding planes from which two specimens were were obtained: (A) parallel to bedding planes; (B) across the bedding planes

Equipment and techniques for triaxial testing

The triaxial test is so important in soil mechanics that it can be considered the standard laboratory test. The main references are Bishop and Henkel's (1962) book and, more recently, Head's (1980). A brief description will be presented here.

Triaxial cell

The basic triaxial cell is presented in figure 8.6. It encompasses an acrylic or perspex transparent cylinder, an aluminum base, a piston and its bushing. The soil specimen is placed on a pedestal that is connected to the outside allowing volume change and pore pressure measurements. Axial load is applied by means of a piston rod and the confining pressure by means of water. Porous stones, covered with filter paper to avoid soil ingress, are placed on the top and the bottom of the specimen to facilitate drainage.

A rubber membrane is placed all around the specimen to avoid contact with cell water. The membrane is sealed on top and bottom by means of rubber bands or o-rings.

Between the specimen and the rubber membrane, a spiral stripe of filter paper is placed, in order to improve drainage and to equalize pore pressures within the specimen and increase the rate of consolidation.

Pore pressures and volume change measurements

Figure 8.7 presents the necessary equipment for pore pressure and volume change measurements in a saturated specimen. They include an electrical transducer, a drainage control valve and a graduated burette.

Drainage control is performed by means of a valve that is the only possible path for incoming or outgoing water. If it is closed, the test is undrained. During

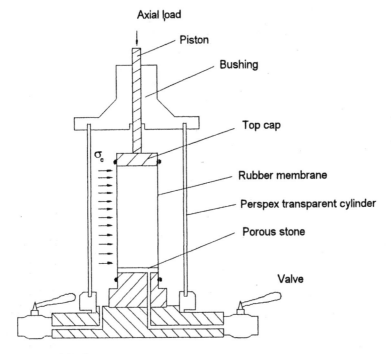

Figure 8.6. The triaxial cell

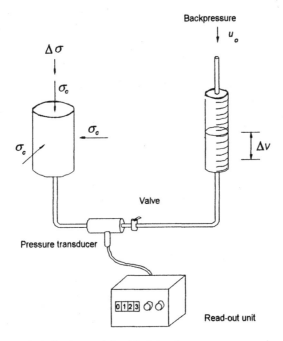

Figure 8.7. Measurement at the base of the triaxial cell: pore pressures or volume change and backpressure.

this test pore pressures can be measured through a pressure transducer. This is a very sensitive instrument having a thin steel diaphragm on which water pressure acts on one side. Its deflection is sensed at the other face by means of strain gauges, which outputs a proportional electrical signal. Using appropriate electronic instrumentation previously calibrated, the pore pressure is displayed in engineering units such as kPa or MPa.

When a saturated specimen is tested in drained conditions, the interest lies in volume change measurements. This is accomplished by measuring the volume of water flowing in or out of the specimen through a calibrated burette (figure 8.7).

Backpressure

Backpressure is a technique for saturating specimens. It is accomplished by applying water pressure u_0 within the specimen (figure 8.7), and at the same time altering the cell pressure σ_{cel} of an equal amount. Therefore, the net confining pressure $\sigma_c = \sigma_{cel} - u_0$ remains unchanged. Backpressuring has no influence on the calculations, since it is equivalent to an increase in the atmospheric pressure. It should be applied in stages, as shown in the example in table 8.1. In most cases a backpressure of 300 kPa is sufficient to ensure specimen saturation.

Backpressure is used in the following cases:

The soil sample was saturated in situ, but due to remoulding during sampling, transportation, storage and trimming may have lost water. Therefore, it is important to ensure initial conditions by means of backpressuring. Air bubbles within the specimen will, then, be dissolved by water pressure and the specimen will become saturated.

The soil is unsaturated in situ, as compacted clay cores used in embankment dams, but may eventually become saturated after reservoir impounding, due to water seepage. If soil testing is aimed at the study of final steady state conditions, it may be desirable to saturate soil specimen by means of backpressuring.

Table 8.1. Example of backpressure stages

Stage	Backpressure u_0 (kPa)	Cell pressure σ_{cel} (kPa)	Confining pressure $\sigma_c = \sigma_{cel} - u_0$ (kPa)
0	0	100	100
1	20	120	100
2	40	140	100
3	80	180	100
•	•	•	•
•	•	•	•
10	300	400	100

In fact, the initial degree of saturation has an important effect on excess pore pressures during shear, as will be discussed later in chapter 11.

Measurements of volume change are easy to perform, but only if the soil is fully saturated. The amount of water that comes in or leaves the soil specimen is observed. If soil is unsaturated, measurements are more complicated, because volume change is not related to the amount of water that gets in or out from the specimen.

Measurements of negative pore pressures: saturated soils which dilate during shear may present a decrease in pore pressure, which would be impossible to measure if it drops below water cavitation pressure. By previously backpressuring the specimen, negative pore pressure changes may be observed, provided the backpressure value is high enough to ensure that the pressure being measured is always positive.

Classification of shear tests according to drainage

Early test classification used the rate of shear as the parameter for classification. Triaxial and direct shear tests could be regarded as *slow* or *quick* (*cf* Lambe, 1951) depending on the rate of shearing. This practice is still used in embankment dam engineering. However, soil engineers realized that the most important aspects in test characteristics are drainage conditions. Shear tests are now regarded as *drained* or *undrained*.

Triaxial tests are generally carried out in two phases The first is the application of an all-around pressure followed (or not) by consolidation, then, the specimen is brought to failure by increasing the deviator stress. Drainage conditions can vary between these stages and the tests are classified as:

Consolidated drained test *CD*

This test comprises the following steps. Drainage valves are initially closed (figure 8.8a). As the confining stress σ_c is applied, pore pressures will vary by an amount Δu. Drainage is permitted and, at the end of consolidation, at the time $t = t^*$, Δu is fully dissipated. The effective isotropic confining stress is σ'_c.

The shear phase starts. Drainage is permitted. The rate of shear is adjusted to allow sufficient time for full pore pressure dissipation. This means high rates in sand, the test being conducted in 15 or 20 minutes. In low permeability soils, like clays, rates of shear are low to allow time for pore pressure dissipation. Shearing may take up to one week in very impervious soils.

Consolidated undrained test *CU*

The consolidation phase is drained (figure 8.8b), as in the previous type. Shear is undrained, and pore pressures are observed throughout the test.

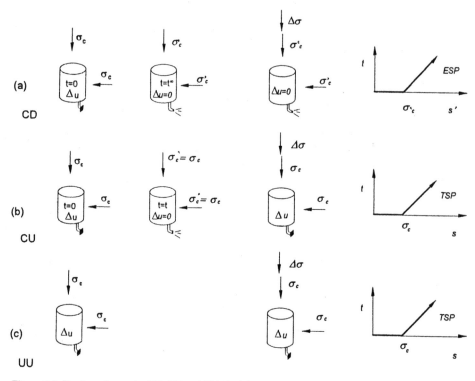

Figure 8.8. Testing phases in *CD*, *CU* and *UU* triaxial tests

Unconsolidated undrained test *UU*

Both test phases (figure 8.8c) are undrained. Pore pressures are usually not measured.

Application of tests according to drainage condition

In the following chapters the purpose of each test type will be studied in more detail.

In sands *CD* tests are used to simulate drained field behaviour. In rapidly cyclic stress conditions, such as in earthquakes, there is no sufficient time for dissipation and pore pressure builds up. This case requires an undrained condition, even for sands.

Clays are subjected to the three types of test depending on the situation to be analysed and on the parameters required.

The direct shear test can only be *CD*, for reasons discussed before.

Test classification according to the stress path at consolidation

During the consolidation phase of triaxial tests the most common situation is the application of an all around confining stress producing an isotropic consolidation. Such tests are known as *CID* or *CIU*, in which *I* stands for isotropic consolidation.

There are special cases of triaxial testing in which the consolidation stresses keep a K_0 relationship. The tests are, then, known as CK_0D and CK_0U.

Most *UU* tests are carried out in isotropic conditions. However, there are a few cases related to embankment dam construction when a very special undrained test is carried out for predicting pore pressure built-up as the embankment rises. This test simulates a constant total stress ratio that occurs in situ and is called constant *K* tests. In this case however, the applicable notation according to this classification, is *UKU*.

Special triaxial tests will be discussed later in chapter 13.

Triaxial test classification according to stress path during shear

In this chapter we have studied only compression tests. This corresponds to the great majority of triaxial tests carried out in practice.

There are special cases in which one is interested in different loading conditions. For example, one may be interested in simulating the unloading that takes place in an excavation, or a lateral compression or extension close to a wall that suffers a lateral displacement. These conditions are studied in special stress path tests. If deviator stress decreases, it is considered an *extension test*. Specimens may be brought to failure in *lateral compression* or *extension*.

Special triaxial test types will be discussed further in chapter 13.

Proposed exercises

8.1. What are the necessary measurements in triaxial testing in order to determine Poisson's ratio of soils?

8.2. Can undrained direct shear tests be conducted? Why?

8.3. Which laboratory test would you select for determining stress-strain parameters of sand: triaxial or direct shear? Explain.

8.4. Dam engineers employ the designation quick test or consolidated-quick test. Do you agree with this classification? Explain.

8.5. What is a backpressure and how is it applied?

Chapter 9

Behaviour of sands

Introduction

Granular materials like sands and gravels are *free draining* soils. Their stress-strain strength properties are, then, determined in drained tests to simulate field behaviour.

There is, however, an exception in which even sands behave as undrained soils. It occurs during rapid cyclic loading in earthquakes. Cycling is so fast that there is no time for dissipation and pore pressures build up can be observed. We know from experience that this occurs in loose fine sands leading to liquefaction. Undrained tests in granular materials will not be discussed here.

In this chapter we will analyse the results of drained triaxial tests in sands, followed by an analysis of direct shear tests. We will learn about stress-strain-strength properties, and factors that affect these properties. The effect of the initial void ratio and confining pressure will be discussed.

The Mohr-Coulomb strength envelope

Three sand specimens were extracted from the same sample and subjected to *CID* triaxial tests. The confining stresses σ'_c were 100, 200 and 300 kPa, respectively. Figure 9.1 shows the test results: a plot of the deviator stress against vertical strain.

Table 9.1 summarizes the results at failure. Confining stress σ'_c, equal to the minor effective principal stress σ'_3, was kept constant throughout the test. The deviator stress at failure $(\sigma_1 - \sigma_3)_f$ was evaluated from the peak of the stress-strain curves, as shown in figure 9.1. The major effective principal stress σ'_{1f} was calculated through the following equation: $\sigma'_{1f} = \sigma'_3 + (\sigma_1 - \sigma_3)_f$.

Data in table 9.1 include the major and minimum effective stresses at failure. The correspondent Mohr's circle of stresses can be drawn, as shown in figure 9.2. A straight line tangent to the circles is also drawn. This line is called

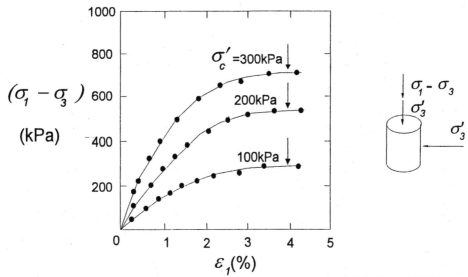

Figure 9.1. Triaxial test results on sand specimens at confining stresses of 100, 200 and 300 kPa

Table 9.1. Data from *CID* triaxial test on sand

Specimen	$\sigma'_c = \sigma'_3$ (kPa)	$(\sigma_1 - \sigma_3)_f$ (kPa)	σ'_{1f} (kPa)
1	100	269	369
2	200	538	738
3	300	707	1007

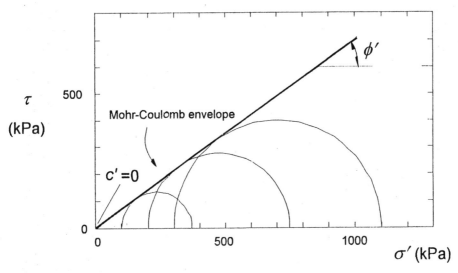

Figure 9.2. Mohr-Coulomb's failure envelope

Mohr-Coulomb's strength envelope. It delimits two regions in the $\tau{:}\sigma$ plot. In the upper one, no state of stress is possible, failure has already occurred. On the other hand, all points below Mohr-Coulomb's strength envelope are possible.

Mohr-Coulomb's strength envelope is represented as a straight line, given by:

$$\tau_{ff} = c' + \sigma'_{ff} \tan\phi' \tag{9.1}$$

where: τ_{ff} is the shear stress in the failure plane at failure; σ'_{ff} is the effective normal stress in the failure plane at failure; c' and ϕ' effective strength parameters, where: c' is the intercept on the ordinate axis, also called the *effective cohesion*; ϕ' is the angle of inclination of Mohr-Coulomb's envelope, also called the *effective friction angle*.

If the sand is *uncemented*, i.e., if it has no grain to grain bond or cement, the shear strength τ_{ff} turns to nil when the effective stresses drop to zero, then: $c' = 0$. This is the reason why sands and gravels are called *cohesionless* soils. Equation 9.1 becomes simply:

$$\tau_{ff} = \sigma'_{ff} \tan\phi' \tag{9.2}$$

Inclination of the failure plane

The theoretical inclination of the failure plane θ_{cr} in a triaxial test can be determined through the pole graphical construction, in chapter 2. Consider the failure plane where stresses σ'_{ff} and τ_{ff} act (figure 9.3a).

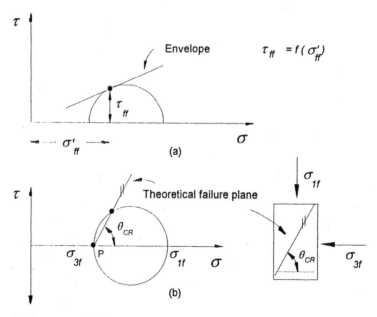

Figure 9.3. (a) Stresses on the failure plane at failure; (b) Theoretical inclination of failure plane obtained graphically through the pole

Starting at the point corresponding to the major principal stress σ'_{1f} (figure 9.3b) a line is drawn parallel to horizontal plane. This line crosses the circle again at pole P that coincides with the location of the minor principal stress σ'_{3f}. From P a line is drawn to the failure point in Mohr's circle. The angle that this line forms with the abscissa is the *theoretical* inclination of the failure plane θ_{cr}, and may be different from actual observations on failed specimens.

Comparison between τ_{ff} and the maximum shear strength τ_{max}

Figure 9.3b shows that the shear strength τ_{ff} occurring at the failure plane is lower than the maximum shear τ_{max}. This occurs when ϕ' value is greater than zero. We obtain $\tau_{ff} = \tau_{max}$ only if Mohr-Coulomb's envelope is horizontal, i.e., $\phi' = 0$.

Transformed strength envelope

Figure 9.4 presents the effective stress paths (data from table 9.1). The horizontal portion of the *ESP's* corresponds to the isotropic compression phase under the confining stress σ'_c. This is followed by the shear phase, corresponding to an increase in the deviator stress and the *ESP*'s are inclined with a 1:1 slope to the right.

Failure occurred at the last points of the *ESP's*, which have coordinates (s_f, t_f). A straight line is fitted through the failure points. This line is the *transformed strength envelope*, given by the following equation:

$$t_f = a' + s'_f \tan\alpha' \tag{9.3}$$

where: a' and α' are effective strength parameters of the transformed strength

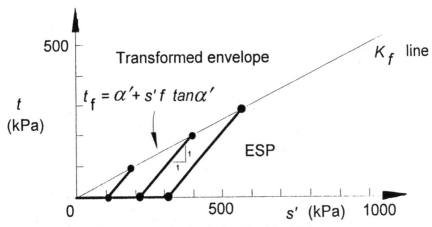

Figure 9.4. The transformed envelope fitted to final points of the *ESP's*

envelope, respectively, the strength intercept and the inclination.

The equivalence between these parameters and the traditional Mohr-Coulomb's c' and ϕ' is deduced in exercise 9.1.

Exercise 9.1

Obtain a relationship between the strength parameters of the transformed envelope and Mohr-Coulomb's envelope.

Solution

Figure 9.5 presents two envelopes for the same Mohr's circle. The coordinate axis $s':t$ coincides with the $\sigma':\tau$ axis. Mohr-Coulomb's envelope touches the circle at point B. The transformed envelope crosses it at point C. It can be verified from elementary geometry that both envelopes intercept each other at A. This point lies on the negative side of the abscissae.

Consider triangles AOC and AOB. They have a common segment AO, and $OC = OB = t_f$, hence:

$$\tan\alpha = t_f / AO \text{ and } \sin\phi = t_f / AO$$

$$\therefore \ \sin\phi = \tan\alpha \tag{9.4}$$

Triangles ADF and ADE have a common side AD, therefore:

$$\tan\alpha = a/AD \text{ and } \tan\phi = c/AD.$$

Replacing AD and introducing equation 9.4:

$$c = \frac{a}{\cos\phi} \tag{9.5}$$

The purpose of the transformed envelope is to enable strength parameters to be obtained by fitting a straight line through experimental data points. Linear re-

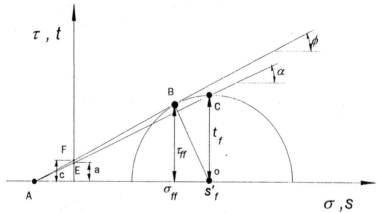

Figure 9.5. Relationship between parameters obtained from the Mohr-Coulomb and the transformed envelope

gression technique can be used. This procedure is easier than drawing a tangent line to three or more Mohr's circles that *do not* have a single tangent. Experimental failure points are never aligned on a unique strength envelope, due to soil lack of homogeneity or experimental difficulties. The parameters a' and α' can, then, be obtained through a linear regression on data points and, afterwards, transformed into Mohr-Coulomb's parameters, through equations 9.4 and 9.5, yielding c' and ϕ'. Exercise 9.2 gives an example.

Exercise 9.2

Given results of *CID* triaxial tests on sand (figure 9.6), obtain: (a) Mohr circles at failure; (b) *ESP's*; (c) Mohr-Coulomb's strength parameters from the σ':τ plot; (d) the same, for the s':t plot; (e) the theoretical inclination of the failure plane with horizontal.

Solution

From data in figure 9.6, the confining pressures and deviator stresses at failure are drawn and presented in table 9.2. Failure was assumed to occur at maximum

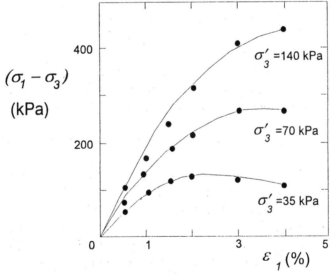

Figure 9.6. Exercise 9.2: Triaxial test results on sand

Table 9.2. *CID* triaxial tests on sand

Specimen	$\sigma'_c = \sigma'_3$ (kPa)	$(\sigma_c - \sigma_3)_f$ (kPa)	σ'_{1f} (kPa)
1	35	69	128
2	70	270	340
3	140	425	565

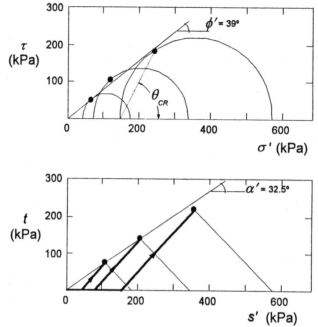

Figure 9.7. Exercise 9.2. (a) Mohr-Coulomb's envelope; (b) Transformed envelope

deviator stress. The major effective principal stress at failure σ'_{1f}, is obtained from the first two columns in table 9.2, through: $\sigma'_{1f} = \sigma'_3 + (\sigma_1 - \sigma_3)_f$.

Figure 9.7a shows Mohr's circles at failure. Mohr-Coulomb's envelope was chosen, considering:

(i) The sand sample is considered uncemented and the cohesion intercept was taken as nil, i.e., $c' = 0$;

(ii) Several Mohr's circles drawn from experimental data *never* have the same tangent. A secant envelope was selected resulting in $\phi' = 40°$.

The theoretical inclination θ_{cr} of failure points is indicated in the figure. The *ESP's* were drawn in figure 9.7b, reaching failure. The transformed envelope was fitted considering $a' = 0$ (i.e., $c' = 0$), giving $\alpha' = 32.5°$. Then, from equation 9.4, $\phi' \cong 40°$, which agrees with the value previously obtained through Mohr's plot.

This exercise shows that it is easier to use the $s':t$ diagram to work out strength parameters from triaxial test data.

The strength envelope at the direct shear test

Mohr-Coulomb's strength envelope can be evaluated from direct shear tests, as shown below in exercise 9.3.

Exercise 9.3

Five identical specimens of sand were taken to failure in the direct shear test. Normal stresses ranged from 0.35 to 1.1 MPa. The results are presented in figure 9.8a. Obtain the Mohr-Coulomb strength envelope and the value of ϕ'.

Solution

Normal stress is assumed to be constant during the test and shear stress increases with lateral displacement of the split box. Failure is assumed to occur at the maximum value, i.e., $\tau_{ff} = \tau_{max}$. Figure 9.8b presents a plot of the shear stress τ against the normal stress σ'_{ff} for each specimen. A straight line envelope is fitted through the experimental data points yielding Mohr-Coulomb's strength parameters.

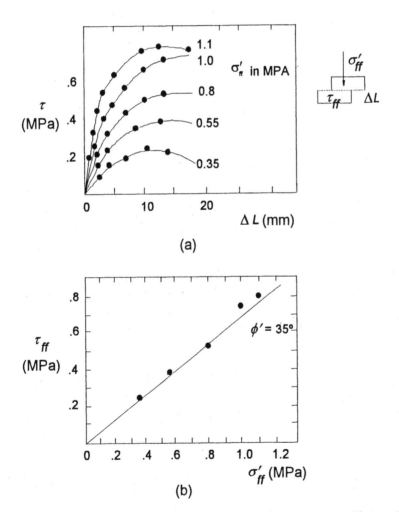

Figure 9.8. Exercise 9.3. Direct shear test results: (a) stress-displacement curves; (b) strength envelope

Factors influencing φ' in sands

The friction angle of sands is influenced by the following factors *slip*, *rolling* and *packing* (figure 9.9). During shear, sand grains can slip against one another, leading to friction. If grains rotate or roll during shear, it will also cause friction.

These factors are, in turn, influenced by shape and surface roughness of sand particles. River sands, for instance, usually present rounded particles and little surface roughness, which tends to decrease internal friction. On the other hand, crushed stone used in concrete present a remarkable surface roughness, which will lead to higher friction.

The third cause of friction is *packing* which is related to grain fabric.

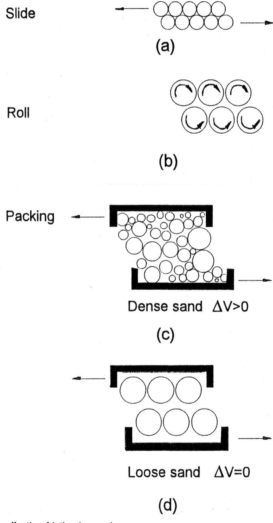

Figure 9.9. Factors affecting friction in sands

Grain fabric

Consider a well-graded sand, presenting a gentle slope on the grain size distribution curve. Imagine that the sample is heavily compacted by vibration to a relative density D_r near 100% and the grains become closely packed together with little space left in between (figure 9.9c). As shear starts, sand particles will tend to ride one over the other, resulting in dilation. The greater the packing, the greater the dilation at shear.

The volume change during shear is called *dilatancy* and has a very important effect on shear strength because a considerable amount of shearing energy is used to cause dilation. On the other hand, if the sand is loose and all particles have the same diameter (figure 9.9d), there is no tendency for dilation. Contraction may occur. Friction is only due to slip and rotation between grains.

Grain packing for one particular sand sample can be evaluated by means of the initial void ratio e_0, or the relative density D_r.

Friction due to grain slipping and rotation depends on the shape and surface roughness, which are intrinsic characteristics of a particular sand. Dilatancy, on the other hand, depends on the state – dense or loose – of the sand. This will be further explored in the following items.

Behaviour under low stress levels

Sand behaviour under low stress level, i.e., up to 300 kPa, will be studied through triaxial test results in figure 9.10 (Taylor, 1948). The data refer to two *CID* test on the same sand, at the same confining stress $\sigma'_c \cong 200$ kPa, but for different initial void ratios. The loose specimen has an initial void ratio $e_0 = 0.834$, and the dense one has $e_0 = 0.605$.

Stress strain curves (figure 9.10a) show very different patterns. The dense sand presents a well-defined peak of strength corresponding to the maximum deviator stress $(\sigma_1 - \sigma_3)_{max}$. The friction angle associated with this value (taking $c' = 0$) is the peak friction angle ϕ'_f. Just after the peak, occurs strain softening, i.e., loss of strength with increasing strain.

The loose sand reaches the maximum deviator stress and thereafter shows plastic behaviour. There is no peak.

The dashed portions of the stress-strain curves were extrapolated because it is generally not possible to continue the test when strains are over 20%. Beyond 20%, the specimens become too deformed and strain measurements are meaningless.

The volumetric strain behaviour is also very different for the loose and the dense sand. The dense specimen dilates as a result of the tight grain packing, but with large strains it tends to stabilization (figure 9.10b). The loose specimen presents a small initial contraction, followed by an equal expansion, and thereafter there is no significant volumetric change.

Figure 9.10c show that both specimens tend to the same void ratio at large strains.

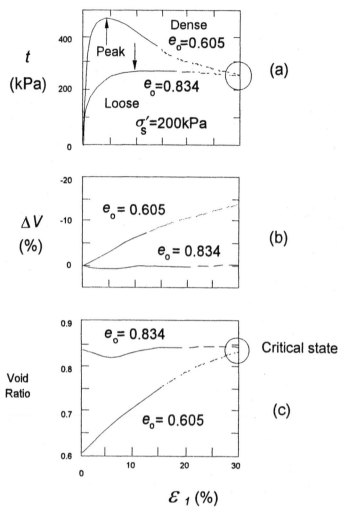

Figure 9.10. Triaxial test results in loose and dense sand at low (200 kPa) confining stress (Taylor, 1948)

Critical state

Both specimens of dense and loose sand (figure 9.10) tend, at large deformation, to a stable state, in which the strength (q or t), the mean effective stress (p' and s') and the void ratio e, do not change. This state has been called *critical* by the soils group of Cambridge University (Schofield and Wroth, 1968, Atkinson and Bransby, 1978). It is mathematically expressed by:

$$\frac{\partial q}{\partial \varepsilon_1} = \frac{\partial p'}{\partial \varepsilon_1} = \frac{\partial e}{\partial \varepsilon_1} = 0 \tag{9.6}$$

or, at the MIT's $s':t:e$ plot, by:

$$\frac{\partial t}{\partial \varepsilon_1} = \frac{\partial s'}{\partial \varepsilon_1} = \frac{\partial e}{\partial \varepsilon_1} = 0 \qquad (9.7)$$

The friction angle at the critical state is ϕ'_{cr} .

The see-saw analogy

Dilatancy and its effect on the shear strength can be studied by means of the simple see-saw analogy created by Rowe (1961 and 1963). Consider a wooden block split into two parts. The split surface in between presents a see-saw pattern as indicated in figure 9.11. If a tangential force is applied to one part of the block, slip failure occurs at the see-saw plane.

Angle ψ , corresponding to the inclination of plane *AA* with the horizontal, controls the volume change behaviour during shear. A loose sand that does not change volume during shear is represented in this model by $\psi = 0$. In this case there is no dilation at shear, the split block will not be displaced upwards.

If ψ is greater than zero, once shear starts, the block will be displaced horizontally and vertically, the latter representing dilation.

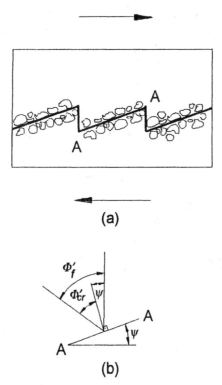

Figure 9.11. See-saw analogy

According to this model, peak friction angle ϕ'_f is a sum of the following components:

$$\phi'_f = \phi'_{cr} + \psi \tag{9.8}$$

where: ϕ'_{cr} is the effective critical state friction angle; ψ is the friction component representing dilatancy.

As an example, consider the results of direct shear tests on Guandu river sand, Rio de Janeiro (Pacheco, 1978), plotted in figure 9.12. The tests were carried out with specimens prepared in different initial void ratios ranging from loose ($e_0 \cong 0.8$) to dense ($e_0 \cong 0.55$). In each test one attempted to obtain the peak friction angle ϕ'_f, and the final value at large strain ϕ'_{cr}. The results were: for the dense sand $\phi'_f \cong 42°$ and $\phi'_{cr} \cong 33°$; for the loose, $\phi'_f \cong 36°$ and $\phi'_{cr} \cong 32°$.

Figure 9.12a shows a best-fit curve through ϕ'_f data and another horizontal line fitted through ϕ'_{cr} data points. They cross at point A with coordinates (e_{cr}, ϕ'_{cr}). The difference $\psi = \phi'_f - \phi'_{cr}$ is plotted in figure 9.12b versus e_0. The

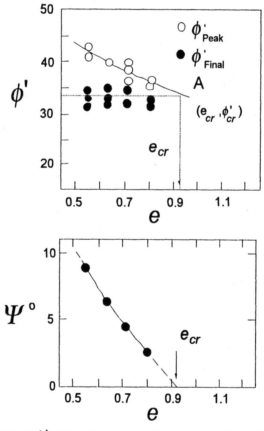

Figure 9.12. Determination of ϕ'_{cr} and e_{cr} for sand from Guandu River, Brazil (Pacheco, 1978)

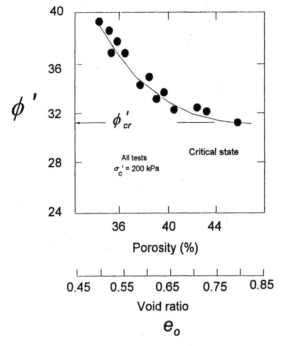

Figure 9.13. ϕ' values from triaxial tests on sand specimens (Rowe, 1961)

void ratio at the critical state e_{cr} can also be drawn from this plot by extrapolation to a $\psi = 0$ condition.

Several *CID* triaxial tests on sand (figure 9.13) were carried out by Rowe (1961) at the same confining stress σ'_c, but varying the initial void ratio e_0. The results show that as e_0 increases, the peak friction angle ϕ'_f tends to the critical state value ϕ'_{cr}.

The following conclusions about the critical friction angle ϕ'_{cr} can be drawn. First, ϕ'_{cr} can be interpreted as a material property that is independent from the state. Second, ϕ'_{cr} is a conservative value for the friction angle, occurring at large strain. If it is used in the assessment of stability, the design will be on the safe side.

Exercise 9.4

For the *CID* triaxial test results in figure 9.10, obtain: the *ESP's*, the peak and critical state strength envelopes, and the values of ϕ'_f and ϕ'_{cr}.

Solution

Test results were summarized in table 9.3, where t_f and t_{cr} are, respectively, t at peak (or failure) and at critical state. The stresses σ'_{1f} and σ'_{1cr} are worked out:

Table 9.3. *CID* triaxial test results for sand

Density	σ'_c	t_f	t_{cr}	σ'_{1f}	σ'_{1cr}
	kPa	kPa	kPa	kPa	kPa
Dense	200	475	250	1150	700
Loose	200	250	250	700	700

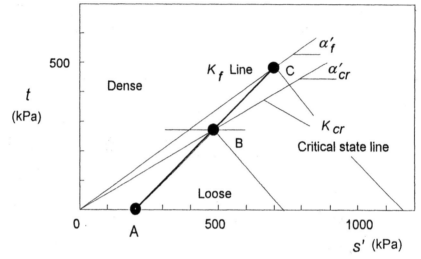

Figure 9.14. Exercise 9.4. *s':t* diagram

The *ESP's* are plotted in figure 9.14. The dense sand starts shear at *A*, ends at *C*, which corresponds to peak or failure, then, returns by the same path to *B*, the critical state. The loose sand starts shear at *A* and goes to *B*, staying at this point until critical state is reached and the test is stopped.

The peak strength envelope (the called K_f line) yields $\alpha'_f \cong 35.5°$. The critical state envelope, or K_{cr} line, yields $\alpha'_{cr} \cong 30°$. Finally, through equation 9.4, the following values are obtained: $\phi'_f = 45°$ and $\phi'_{cr} = 35°$.

Influence of stress level

The influence of stress level will be discussed utilizing data from Lee (1965), also published by Holtz and Kovacs (1981). These data refers to *CID* triaxial testing on dense sand specimens at several confining stresses σ'_c, ranging from 100 kPa to extremely high values like 13.7 MPa (figure 9.15). Stress-strain curves (figure 9.15a) were normalized, i.e., divided by σ'_3.

At low confining stresses, stress-strain curves show a well-defined peak and dilation (figure 9.15b) occurs. As confining stresses increase, the peaks tend to disappear. The initial slope of the stress-strain curves also decreases. The vol-

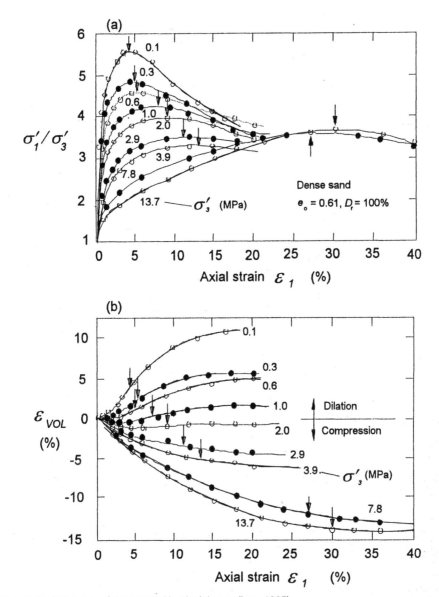

Figure 9.15. Behaviour of dense sand in triaxial tests (Lee, 1965)

ume change behaviour changes from dilation, at low stresses, to contraction at higher stresses.

Figure 9.16 presents plots of the loose specimens. As confining stresses increase, the stress-strain curves (figure 9.16a) indicate a decrease in the initial slope and an increase in the peak strain. Volumetric strains at low confining stress (figure 9.16b) show a slight compression followed by dilation. At high confining stresses, volume compression is high.

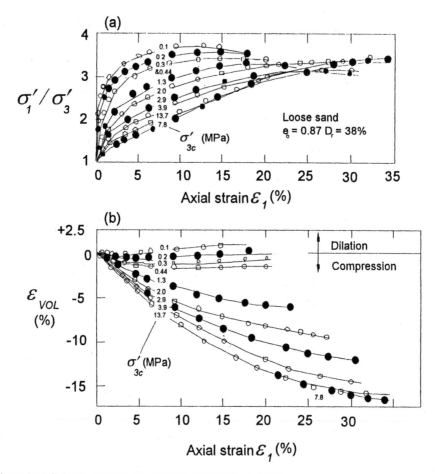

Figure 9.16. Behaviour of loose sand in triaxial tests (Lee, 1965)

The effect of high confining stresses on both loose and dense specimens is explained by grain crushing which affects the volume change behaviour.

Non-linearity in the strength envelope

In a wide range of normal stresses all soils will present a curvature in their shear strength envelope, as shown in figure 9.17. Mohr-Coulomb's straight line envelope is no longer directly applicable. In granular materials, this can be caused by cementation between grains, which is destroyed during shear at higher stress levels, a change in soil density, or grain crushing at high stresses.

Calcareous sands frequently present cementation and very weak grains. As stress level increases, the grains can be easily broken during shear (Ortigao et al., 1985). Silica and quartz sands have much stronger grains and typically exhibit no cementation.

For a wide range of stress levels, as shown in figure 9.17, the strength envelope may be curved. Mohr-Coulomb's straight line assumption is only an approximation and is restricted to a limited range of stresses. This emphasizes the need for testing soils at the expected stress level that will occur during construction.

Coarse granular materials, such as rockfill employed in dams and breakwaters present a significant curvature in the strength envelope (Barton and Kjaernskli, 1981, Charles and Soares, 1984). When analysing the stability of high dams, especially those higher than 60 m, this phenomenon plays an important role and should be taken into account. The same occurs in the analysis of high slopes of fractured rocks (Hoek, 1983).

The non-linearity in the strength envelope can be taken into account in the following ways. First, by adopting small linear portions along the envelope, each one having different c' and ϕ' values as a function of the stress level. Then, selecting c' and ϕ' to be used in the analysis. This method presents the disadvantage of using two parameters as a function of the stress level and a fictitious value of cohesion for a granular material.

The second and the preferred method is to take c' always equal to zero and to vary ϕ' with the stress level. ϕ' is tangent to *one* single Mohr's circle. It is assumed to vary with the stress level, as indicated in figure 9.17. In this example, a higher ϕ' of 47° is obtained for the low stress range of $\sigma'_{ff} = 0$-0.3 MPa. A low ϕ' of 31° is obtained for higher stresses (σ'_{ff} up to 2 MPa).

The following equations have been employed to express the variation of ϕ' with stress range:

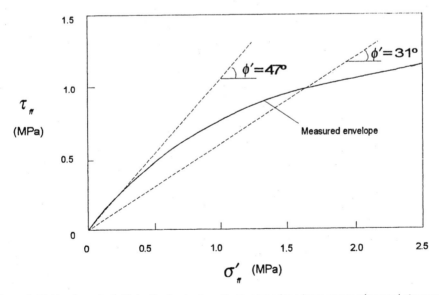

Figure 9.17. Non-linearity in Mohr-Coulomb strength envelope for a large range of normal stresses

Wong and Duncan (1974) proposed a $\phi' = f(\sigma'_{ff})$ type function to express variation of ϕ' with the stress level granular materials in dams:

$$\phi' = \phi'_0 - \Delta\phi'\log\left(\frac{\sigma'_c}{p_a}\right) \tag{9.9}$$

where: ϕ'_0 is the ϕ' value corresponding to σ'_c equal the atmospheric pressure p_a. The parameter $\Delta\phi'$ corresponds to a reduction in ϕ' which occurs in a log cycle (10 times) of σ'_c.

Mello (1977) proposed, for rockfills, the exponential equation below:

$$\tau_{ff} = A\left(\sigma'_{ff}\right)^b \tag{9.10}$$

where: A and b are parameters determined by curve fitting to experimental data.

Critical state line

If a soil reaches the critical state, it will be located on the K_{cr} line in the $s':t$ plot. Let's look now what happens to the same soil in the $s':e$ space. Consider the triaxial stress data by Lee (1965) shown in figures 9.15 and 9.16. For this sand, *CID* triaxial tests were carried out keeping σ'_3 constant, but varying the initial void ratio e_0 in each specimen. Plotting a similar graph as that in figure 9.12b, the void ratio at the critical state e_{cr} was obtained for each σ'_3. Since σ'_3 is constant, then: $s'_0 = \sigma'_3$. The pairs of points (s'_0, e_{cr}) are plotted in figure 9.18 in which the scale of stresses is logarithmic, which yields an approximate linear relationship.

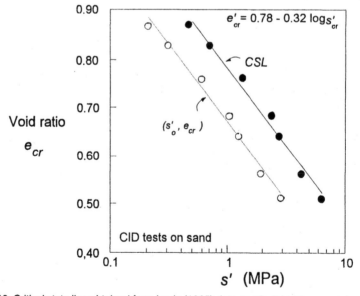

Figure 9.18. Critical state line obtained from Lee's (1965) data on triaxial tests on sand

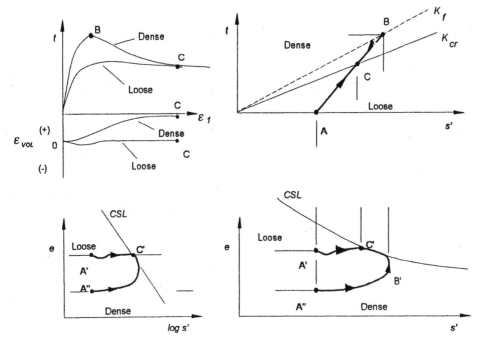

Figure 9.19. Comparison between the behaviour of loose and dense sand under low and high levels of confining stresses: (a) stress-strain and volume change curves; (b) *ESP*'s in the *s':t* diagram; (c) *s':e* diagram; (d) log *s':e* diagram

The next step was to work out the mean stress s'_{cr} at the critical state value corresponding to each e_{cr}. This was done through the following equation:

$$s'_{cr} = 0.5\, s'_0 \left[1 + (\sigma'_1/\sigma'_3)_{cr}\right]$$

where: $(\sigma'_1/\sigma'_3)_{cr}$ is the principal stress ratio at critical state, which was estimated as 3.5 from figures 9.15a and 9.16a.

The resulting points, with coordinates (s'_{cr}, e_{cr}), were plotted in figure 9.18, where they fall in a single *Critical State Line (CSL)*.

These experimental data suggest a linear relationship for points in the critical state in both spaces *s':t* and *s':e*, the latter being a log function.

The assumed behaviour in *CID* triaxial tests is summarized in figure 9.19. Two specimens, a dense one and a loose one, were tested and the stress-strain curves compared in figure 9.19a. The results are similar to the tests presented before in figure 9.10. The *ESP's* are shown in figure 9.19b. They start at point *A*. Point *C*, corresponds to the critical state in both specimens and *B* is the peak of the *ESP* for the dense sand. Figures 9.19c and 9.19d present the behaviour in the *s':e* and log *s':e* plots. The loose sand begins shear at *A'* and reaches the critical state at *C'*. The dense one, begins at *A''*, reaches the peak at *B'*, and its volume keeps increasing until it reaches the critical state at *C'*.

In the log $s':e$ plot, all points at the critical state are assumed to fall in a straight line: the *CSL*.

Prediction of ϕ' as a function of density and stress level

Bolton (1986) proposed a semi-empirical method for the prediction of ϕ' as a function of the relative density D_r (given by equation 1.1) and the mean stress level p'. In this method, ϕ' is given by equation 9.8, and ψ is given by the empirical equation:

$$\psi = n D_{rc} \tag{9.11}$$

where: n is an empirical parameter equal to 5 for axisymmetrical conditions, and 3 for plane strain. D_{rc} is the corrected relative density given by the following empirical equation:

$$D_{rc} = D_r \left(10 - \ln p'\right) - 1 \tag{9.12}$$

Equation 9.12 is only valid for D_{rc} between 0 and 4. The mean effective stress p' is given in kPa.

Exercise 9.5

Use Bolton's method to estimate the change in ϕ'_f for two sand specimens submitted to *CID* triaxial test under confining stresses of 20 kPa and 1 MPa. Given: $D_r = 70\%$, $\phi'_{cr} = 35°$.

Solution
D_r is calculated by means of equation 9.12 for values of $p' = 20$ kPa and 1 MPa:

p' (kPa)	D_{rc}
20	$D_{rc} = 0.70$ (10 - ln 20) - 1 = 3.9
1000	$D_{rc} = 0.70$ (10 - ln 1000) - 1 = 1.2

Since D_{rc} values lie within the 0-4 range, Bolton's method is applicable. Values of ψ and ϕ'_f are given in the table below, from equations 9.11 and 9.8:

p' (kPa)	ψ (degrees)	ϕ'_f (degrees)
20	5 x 3.9 = 19.5	35 + 19.5 = 54.5
1000	5 x 1.2 = 6	35 + 6 = 41

Typical values of ϕ'

At the preliminary stages of a project geotechnical engineers may estimate stability or bearing capacity based only on visual inspection of soils. Table 9.4 and figure 9.20 can be valuable for obtaining a preliminary estimation of ϕ' of cohesionless soils.

The values above should be reduced if *mica* is detected. Mica is a mineral that presents a very low friction angle, of the order of only 10°. It strongly decreases the shear strength of silts and sands, even if only small traces of mica are observed. Therefore, it is extremely important to bear this in mind during field inspections, mainly in residual soils from granite and gneiss, in which mica is frequently present.

Another cause for the reduction of ϕ' is the grain shape. River sand and

Table 9.4. Typical values of ϕ' (degrees)

Material	Dense	Medium dense	Critical state
Silt	30 - 34	28 - 32	26 - 34
Fine to medium sand	32 - 36	30 - 34	26 - 30
Well graded sand	38 - 46	34 - 40	30 - 34
Sand and gravel mixture	40 - 48	36 - 42	32 - 36

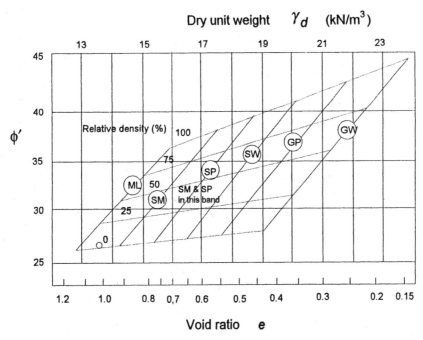

Figure 9.20. Chart for estimating ϕ' in sands, silts and gravels (Navfac DM-7, 1971)

gravel deposits present rounded grains and should have ϕ' reduced in relation to the values portrayed in table 9.4.

Figure 9.20 enables the selection of ϕ' as a function of the dry unit weight γ_d or the initial void ratio e_0 and the relative density D_r. ϕ' is chosen as a function of the soil classification type, according to the *USCS* system described in chapter 1.

Proposed exercises

9.1. Define critical state and present the mathematical equation describing this condition.

9.2. Explain see-saw analogy.

9.3. Consider a direct shear test in which the normal stress is 100 kPa and the shear stress at failure is 35 kPa. Assuming $c' = 0$, obtain ϕ'. Is it possible to work out deformation moduli from this test? Why?

9.4. A direct shear test on sand presented the data in the table below. The shear box dimensions were 75 mm x 75 mm cross section and 10 mm height. The vertical load was 2.4 kN. Obtain the shear stress at the failure plane τ and plot the following diagrams: τ x horizontal displacement; vertical displacement x horizontal displacement. Take $c' = 0$ and obtain ϕ'.

Horizontal displacement	Vertical displacement	Lateral load
(mm)	(mm)	(N)
8.89	3.56	0
8.82	3.54	356
8.63	3.52	721
8.44	3.51	1014
7.92	3.53	1428
7.18	3.59	1655
6.38	3.63	1770
5.49	3.65	1744

9.5. The results of two *CID* triaxial tests on two specimens of the same sand, at the same initial void ratio $e_0 = 0.65$, are presented in the following table. Obtain: (a) Plots of $t:\varepsilon_1$, $s':t:e$ and $\varepsilon_{vol}:\varepsilon_v$; (b) Young's modulus and Poisson's ratio at the beginning of the test and at a stress level of 50% (at a stress 50% less than the maximum deviator stress); (c) friction angles at failure and at the critical state; (d) Mohr's circles at failure and (e) the theoretical inclination of the failure plane.

Specimen 1			Specimen 2		
$\sigma_3 = 100$ kPa			$\sigma_3 = 3$ MPa		
ε_1	$(\sigma_1 - \sigma_3)$	ε_{vol}	ε_1	$(\sigma_1 - \sigma_3)$	ε_{vol}
(%)	(kPa)	(%)	(%)	(kPa)	(%)
0	0	0	0	0	0
1.71	325	-0.1	0.82	2090	-0.68
3.22	414	0.6	2.5	4290	-1.8
4.76	441	1.66	4.24	5810	-2.71
6.51	439	2.94	6	6950	-3.36
8.44	405	4.1	7.76	7760	-3.38
10.4	370	5.1	9.56	8350	-4.27
12.3	344	5.77	11.4	8710	-4.53
14.3	333	6.33	13.2	8980	-4.71
16.3	319	6.7	14.9	9120	-4.84
18.3	318	7.04	16.8	9140	-4.92
20.4	308	7.34	18.6	9100	-4.96
			20.5	9090	-5.01

9.6. Estimate ϕ' for: (a) well-graded sand gravel with $\gamma = 20$ kN/m^3; (b) poorly graded silty sand with $\gamma = 15.5$ kN/m^3 ; (c) poorly graded gravel with in situ void ratio of 0.4.

Chapter 10

Drained behaviour of clays

Introduction

This chapter deals with the behaviour of clay in *CID* drained triaxial tests. The rate of loading in drained testing of clays is slow to ensure full pore pressure dissipation.

Apparently clays are a totally different material from sands. We shall see, however, that there are many similarities, and that the same critical state model can also be adopted.

Test phases

CID triaxial test phases are sketched in figure 10.1, as studied before in chapter 8. At the beginning of the test a confining pressure σ_c is applied (figure 10.1a), leading to an excess pore pressure Δu within the specimen. Drainage valves are opened (figure 10.1b) and consolidation is allowed until full dissipation of Δu. The duration of this phase is 24 to 48 hours in most cases. At the end of consolidation the specimen has changed its volume and the excess pore pressure is zero.

Keeping drainage valves opened, the deviator stress $(\sigma_1 - \sigma_3)$ is applied at a controlled rate to ensure that pore pressures are negligible during the test. In very impervious clays, this implies adopting very small rates, which may lead to a test duration up to one week. Volume change is recorded during the test.

Behaviour of normally consolidated clays

The behaviour of normally consolidated clays will be studied through typical *CID* triaxial test results of the Rio de Janeiro clay presented in figure 10.2. Three specimens were tested under the confining pressures of 70, 200 and 700 kPa. Stress-strain curves and volumetric versus axial strains are shown.

Like sands, the strength of each specimen, i.e., the maximum value of the

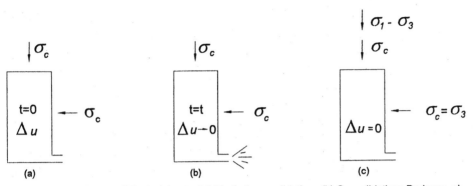

Figure 10.1. Phases in *CID* triaxial test. (a) Start of consolidation; (b) Consolidation: Drainage allowed; (c) Drained shear

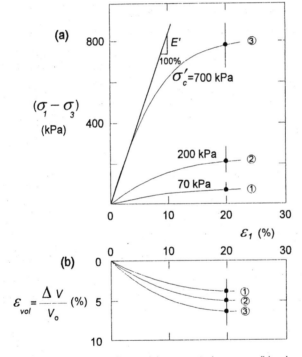

Figure 10.2. Results of *CID* test in *NC* clay: (a) stress-strain curves; (b) volume change versus axial strain curves

deviator stress $(\sigma_1 - \sigma_3)_{max}$, increases with confining stress. Once a maximum value is reached it does not vary significantly until the end of the test.

Young's modulus E' in drained conditions can be evaluated from the slope of the initial part of the stress-strain curve as shown in figure 10.2a. A different value of E' is obtained for each value of the confining stress.

Compressive volumetric strains are observed. They increase steadily and tend to reach their maximum for a corresponding value of axial strain of approximately 20%. Beyond this, the volume of the specimens does not change significantly.

Critical state

It is possible to define for the results in figure 10.2 a stable state at large strains in which volumetric strains and strength do not change any longer. Values of s', t (or p' and q) and e, thereafter, remain constant. This is called critical state and equation 9.6 applies.

Mohr-Coulomb's strength envelope

Results shown in figure 10.2a have been replotted in figure 10.3a. Values of t_{max} are indicated and correspond to failure and to critical state, i.e., $t_{cr} = t_f = t_{max}$.

Figure 10.3b presents the *ESP's* during shear. They begin on the s' axis, at $s' = \sigma'_c$, and end at a point with the following coordinates: (t_{max}, s_{max}).

The transformed envelope is obtained by interpolation through the points (t_{max}, s_{max}), resulting in a straight line intercepting the origin, yielding the fol-

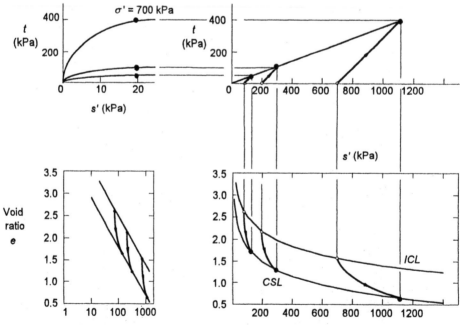

Figure 10.3. Results of *CID* test in *NC* clay: (a) stress-strain curves; (b) *s':t* diagram and *ESP's*, (c) *s':e* diagram; (d) log *s':e* diagram

lowing parameters: $a' = 0$ and $\alpha' = 22.5°$, corresponding to $c' = 0$ and $\phi' \cong 25°$. Therefore, the drained strength of *NC* clays can be described by Mohr-Coulomb's equation 9.2, $\tau_{ff} = \sigma'_{ff} \tan\phi'$, also used for sands.

Since the failure condition coincides with the critical state condition, lines K_f and K_{cr} are coincident (figure 10.3b).

Isotropic consolidation line *ICL* and critical state line *CSL*

The Rio de Janeiro clay specimens were initially consolidated back to the virgin line or, as shown in figures 10.3c and 10.3d, to the isotropic consolidation line *ICL*. These figures indicate the paths in spaces *log s':e* or *s':e* from the start of shear up to the critical state. The critical state conditions in the *s':e* plot occur on an unique line called critical state line *CSL*.

The following equations (figure 10.4) describe the *ICL* and *CSL*:

$$ICL: \quad e = e_{co} - C_c \log s'$$

$$CSL: \quad e = e_{cs} - C_c \log s'$$

where: e_{co} and e_{cs} are void ratios in which $s' = 1$ kPa.

Normalised behaviour

The concept of normalised behaviour (Ladd and Foott, 1974) is presented in figure 10.5. It evolves from empirical observations that, for the great majority of fine grained soils at the same *OCR*, stress-strain-strength properties are a unique function of the confining stress.

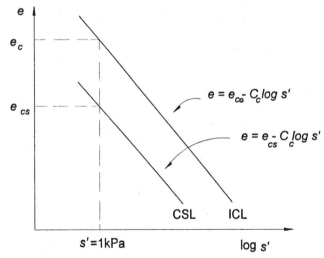

Figure 10.4. *CSL* and *ICL*

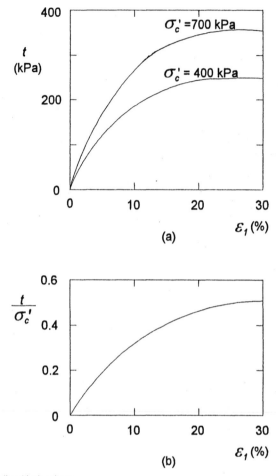

Figure 10.5. Normalised behaviour

As an example, consider the stress-strain curves presented in figure 10.5a for two specimens of the same *NC* clay consolidated at different confining stresses of 400 and 700 kPa, respectively. Figure 10.5b presents the same data, but the ordinates have been normalised, i.e., divided by the confining stress σ'_c. The clay is considered to present normalised behaviour because the curves are coincident in the normalised plot. In practice, they would fall into a narrow range.

Using this concept for a *CID* test on the Rio de Janeiro clay, previous data were replotted in figure 10.6. The stress-stress curves and the *ESP's* coincide, indicating that this concept applies. The critical state corresponds to just one point in the normalised $s':t$ diagram.

Experience in applying this concept to clayey soils indicates that cementation between particles or high sensitivity are conditions under which normalised behaviour cannot be applied.

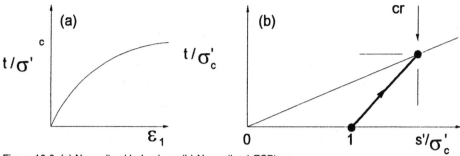

Figure 10.6. (a) Normalised behaviour; (b) Normalised *ESP's*

Exercise 10.1

Predict the behaviour of a *NC* clay in a *CID* triaxial test. The sample is isotropically consolidated from 40 to 100 kPa, and then sheared at a constant confining stress $\sigma'_c = 100$ kPa. Given: $\phi' = 25°$, $C_c = 2.01$, $e_{co} = 5.72$ and $e_{cs} = 5.70$.

Solution

The *s':t:e* diagram is plotted in figure 10.7. The *ESP* in the *s':t* diagram (figure 10.7a) is plotted considering that the isotropic consolidation corresponds to the *AB* portion. Point *A* has coordinates $t_A = 0$ and $s'_B = \sigma'_{cA} = 40$ kPa and point *B*, $t_B = 0$ and $s'_B = \sigma'_{cB} = 100$ kPa.

The *ESP* in the shear phase presents a 1:1 slope and ends at the critical state at *C*, which lies on the K_{cr} line (slope $\alpha' \cong 22.9°$, calculated from the given ϕ'). The coordinates of point *C* are: $s'_{cr} = 173.2$ kPa, obtained graphically, and $t_{cr} = s'_{cr} \tan\alpha' = 173.27 \times \tan 22.9° = 73$ kPa.

The *ICL* and *CSL* lines are plotted point by point through their equations:

$$ICL: \ e = 5.72 - 2.01 \ \log s'$$

$$CSL: \ e = 5.70 - 2.01 \ \log s'$$

These equations allows the calculation of the void ratios at the end of consolidation and the final critical state value. Results are summarized in the table below and plotted in figure 10.7:

Coordinates		Isotropic consolidation		Critical state
		from Point *A*	to Point *B*	Point *C*
s'	(kPa)	40	100	173.2
t	(kPa)	0	0	73
e		2.5	1.7	1.2

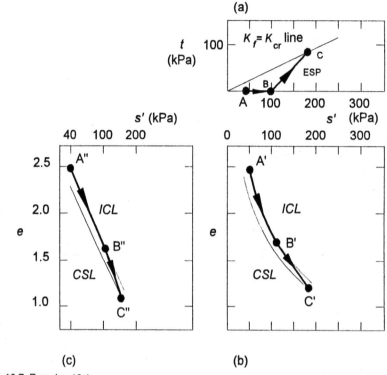

Figure 10.7. Exercise 10.1

Correlations for obtaining ϕ'

The value of ϕ' for normally consolidated clays can be estimated from empirical correlations as those presented in table 10.1, in which ϕ' is related to simple test results, like the Atterberg limits. Figure 10.8 presents data used by Kenney (1959) and other researchers.

Exercise 10.2

Predict ϕ' for the Rio de Janeiro soft clay in which *PI* and *LL* are, respectively, 80 and 150%.

Solution

Applying correlations from table 10.1:

Kenney: $\phi' = \text{arc sin} \ (0.82 - 0.24 \log 80) = 21.3°$

Mayne: $\phi' = \text{arc sin} \ (0.656 - 0.409 \dfrac{80}{150}) = 26°$

Table 10.1. Correlations of ϕ' for *NC* clays

Equation	Reference
$\sin \phi' = 0.82 - 0.24 \log PI$	Kenney (1959)
$\sin \phi' = 0.656 - 0.409 \dfrac{PI}{LL}$	Mayne (1980)

where: *PI* = plasticity index; *LL* = liquid limit

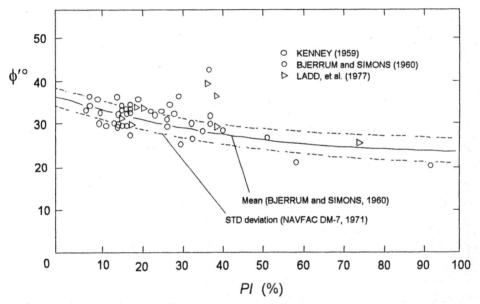

Figure 10.8. Correlation between ϕ' and *PI* for *NC* clays (Holtz and Kovacs, 1981)

Isotropic consolidation and overconsolidation

The behaviour of a *NC* clay specimen during isotropic consolidation in a triaxial cell is presented in figure 10.9. Data points will plot on the *ICL*.

Overconsolidation can be simulated in the laboratory by reducing the confining stress and allowing the specimen to swell. This can be represented in the *s':e* diagram as shown in figure 10.9 for similar specimens consolidated under different confining stresses. During unloading, the specimens will follow the swelling line *SL*, as discussed in chapter 6. The *OCR* is obtained as the ratio between the maximum effective consolidation stress and the pre-shear consolidation stress. If the stress reduction is sufficiently big, corresponding to an *OCR* greater than 8, the swelling line may intercept the critical state line in the *s':e* plot.

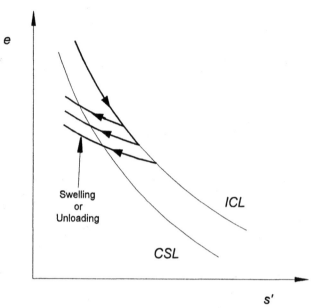

Figure 10.9. Swelling or unloading

Behaviour of overconsolidated clay

The effect of overconsolidation will be analysed by means of the results published by several researchers (e.g., Henkel, 1960, Bishop and Henkel, 1962, and Ladd, 1971).

We will compare the behaviour of two samples of the same clay. Both were consolidated at the same confining stress σ'_c, becoming equally normally consolidated. Then, the confining stress was reduced in one of them, and it was allowed to swell, becoming highly overconsolidated with an *OCR* over 15. Both samples were, then, brought to failure in drained triaxial tests. The results are summarized in figure 10.10, in which all stresses were normalized by the laboratory overconsolidation stress σ'_{vm}. In these tests σ'_{vm} is equal to the initial confining stress σ'_c.

The *NC* clay, as commented before in this chapter, does not show a peak in its stress-strain curve and experiences volumetric compression. Critical state is reached at an axial strain of 20%.

The *OC* clay presents a peak in its stress-strain curve followed by strain softening, i.e., strength reduction with increasing strain. Volumetric deformation shows a slight decrease, soon recovered, and then increases again until the end of the test, at an axial strain of 20%. The test was halted at this strain due to the difficulty, discussed in chapter 8, of measuring large strains in the triaxial test. Beyond a 20% strain, shape distortions of specimens are too large and strain measurements become meaningless. As a result, the conclusions regarding large

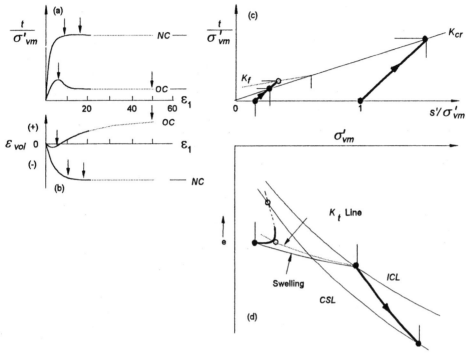

Figure 10.10. Comparison between the behaviour of *NC* and *OC* clays

deformation behaviour were drawn from other tests rather than triaxial ones.

For the overconsolidated clay it is assumed that at large strains the volume no longer varies, enabling the sample to reach critical state conditions.

The *ESP's* are compared in figure 10.10c. The *NC* clay is brought to failure and to critical state at the K_{cr} line, which is coincident with the K_f line for this clay. The *ESP* for the *OC* clay crosses the K_{cr} line, reaches the K_f line and finally drops, returning to the K_{cr} line again.

In *OC* clays failure envelope K_f lies above the critical state line. The intercept of the K_f line with the ordinates results in an effective cohesion c'. At large strains, c' may decrease as the *ESP* returns to the critical state envelope.

The $s':e$ diagram in figure 10.10d shows that both samples were initially consolidated to the same point on the *ICL*. Then, the *OC* clay was allowed to swell at lower stresses and followed the swelling line, crossing the *CSL* and reaching a point at the left of this line. During shear, the *NC* sample decreases its volume until it reaches the *CSL*. The *OC* clay, on the other hand, increases its volume as it moves to the *CSL*.

At this point, we studied the difference in the behaviour of an *NC* and an *OC* clay with a high *OCR*. What will be the effect of intermediate and small values of *OCR's*?

Consider 4 specimens of the same clay at different *OCR's*. The test data are presented in figure 10.11 and in table 10.2. All specimens were isotropically consolidated up to point *A'*, along the *ICL* (figure 10.11b). The *NC* sample remained at this point, while the others were swelled back to points *C'*, *E'* and *H'*. After consolidation they were tested in drained shear.

The *NC* specimen starts shear at *A'* and reaches the critical state at *B*. The

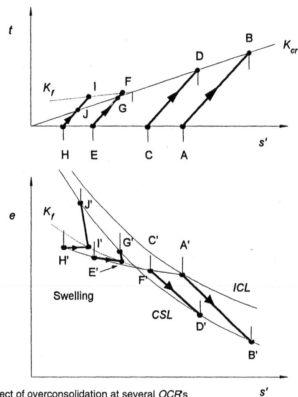

Figure 10.11. Effect of overconsolidation at several *OCR*s

Table 10.2. Consolidation and shear *ESP's* for the tests shown in figure 10.11

Test phase		NC	OC	OC	OC
		Stress path			
Isotropic conslolidation		to A	to A	to A	to A
		to A'	to A'	to A'	to A'
Swelling back simulating overconsolidation		--	AC	AE	AF
		--	A'C'	A'E'	A'F'
State before shear		NC	OC	OC	OC
Drained shear		AB	CD	EFG	HIJ
Critical state	K_{cr}	B	D	G	J
	CSL	B'	D'	G'	J'

slightly overconsolidated sample, starts at C and reaches the critical state at D. In both cases, a decrease in volume was recorded during shear.

The specimens starting shear at H' and E, at the left of the CSL, presented a different behaviour. The volume increased towards points J' and G'. The $ESP's$ travelled upwards beyond the K_f line, but eventually tended to return to the critical state envelope.

Exercise 10.3

An NC clay sample was isotropically consolidated under $\sigma'_c = 50$ kPa giving $e_o = 3.8$ at end of consolidation (point A_1, figure 10.12). Then, it was further consolidated to $\sigma'_c = 1000$ kPa (point B) followed by swelling under $\sigma'_c = 50$ kPa (point A_2). It was, then, sheared in drained compression. Assuming that the ESP reaches the critical state at the end of the test (point C), plot the the $s':t:e$ diagram. Given the critical state parameters: $\phi' = 42.4°$ ($\alpha' = 34°$), $C_c = 1$, $C_s = 0.083$, $e_{co} = 5.5$ and $e_{cs} = 5$.

Solution

The equations of the ICL, CSL and K_{cr} lines are:

$$ISL: \quad e = 5.5 - \log s'$$

$$CSL: \quad e = 5 - \log s'$$

$$K_{cr}: \quad t = s' \tan 34°$$

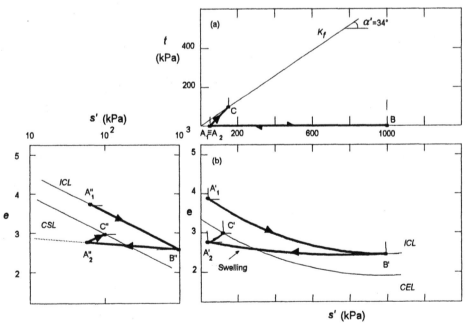

Figure 10.12. Exercise 10.3

They are plotted in figures 10.12a, 10.12b and 10.12c. The following *ESP's:* consolidation A_1B, swelling BA_2 and shear A_2C were obtained from the equations above. The void ratio at the end of the swelling phase (BA_2) was determined from point B, considering a swelling line with a $-C_s$ slope. Data are summarized below:

Point	s'	t	σ'_v	σ'_h	e
	kPa	kPa	kPa	kPa	
A_1	50	0	50	50	3.8
B	1000	0	1000	1000	2.5
A_2	50	0	50	50	2.75
C	150	100	250	50	3.0

Regions in *s':t:e* space

It is convenient to divide the *s':t:e* space in the following regions (figure 10.13):

– POSSIBLE: Possible and stable states of stress are those below the K_{cr} and the K_f lines.

– IMPOSSIBLE: All stress states above the K_{cr} and K_f lines are impossible, because failure has already occurred. In the *s':e* plot, the state of stress at the right of the *ICL* is also impossible because this region corresponds to materials still under sedimentation. Soils can only occur at the left of the *ICL*.

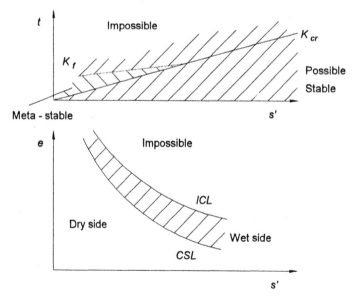

Figure 10.13. Regions in the *s':t:e* space

– META-STABLE: Is the region between the K_{cr} and K_f lines, correspond-ing to *ESP's* that go beyond the critical state envelope but may return to the K_{cr} line.

– WET SIDE: Is the region between the *ICL* and the *CSL*, corresponding to the region in which all samples sheared by drained compression decrease in vol-ume.

– DRY SIDE: Is the region at the left of the *CSL*, where lie the overconsoli-dated samples that increase in volume during drained compression.

Use of drained strength in stability analysis

Clays can only be treated in drained conditions if loading is very slow, and enough time is allowed for full pore pressure dissipation. Then, Mohr-Coulomb's drained shear strength equation $\tau_{ff} = c' + \sigma'_{ff} \tan \phi'$ and effective stress parameters c' and ϕ', can be used for stability calculations.

Some practical applications of drained strength in stability analysis are shown in figures 10.14 (Ladd, 1971) and 10.15. The first case refers to an em-

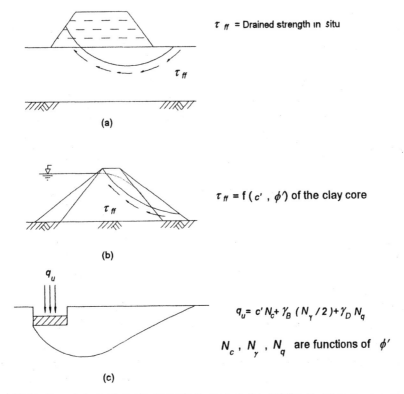

Figure 10.14. Use of drained shear strength to analyse the stability of: (a) embankment on soft ground constructed very slowly; (b) Clay core in a earth dam long time after reservoir filling; (c) shallow foundation slowly loaded (Ladd, 1971)

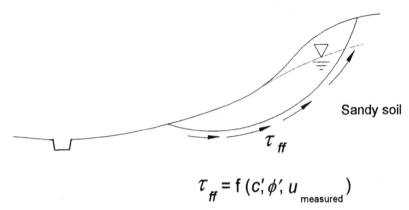

$$\tau_{ff} = f(c', \phi', u_{\text{measured}})$$

Figure 10.15. Slope stability in saprolitic residual soil

bankment placed very slowly on a soft foundation (figure 10.14a), which experiences no excess pore pressure during loadind, and the foundation soils can be assumed as fully drained. The mobilized strength along the likely slip surface is, therefore, the drained strength of the clay.

Figure 10.14b shows a dam after many years of reservoir impoundment when a steady state of seepage has occurred. The mobilized strength in the clay core is also the drained strength.

A shallow foundation on clay many years after construction is shown in figure 10.14c. Drained strength parameters are employed in order to assess the bearing capacity in drained conditions.

Figure 10.15 presents a slope on silty sand. The mobilized strength along the failure surface is the drained strength, as a function of c', ϕ' and pore pressure u.

Comparison between the drained behaviour between clays and sands

As a final step in this study of the drained behaviour of clays, let's compare clays and sands *CID* triaxial compression. Dense sands at low confining stresses present a sharp peak in the stress-strain curve followed by softening. Dilation is observed during shear. A similar behaviour is observed in overconsolidated clays, which begin shear from the dry side.

Loose sands and clays sheared from the wet side present similarities: both do not show a peak strength.

In summary, despite the strong differences, it is possible to find similarities in the behaviour of clay and sands, moreover, the same theoretical framework can be used to model the behaviour of such different materials.

Proposed exercises

10.1. Define: (a) critical state in drained conditions; (b) *CSL*; (c) *ICL*.

10.2. An *NC* clay specimen having the following parameters $\phi' = 33°$, $C_c = 1.2$, $C_s = 0.02$, was isotropically consolidated under $\sigma'_c = 50$ kPa giving $e = 3.1$ at the end of consolidation. Then, it was sheared in triaxial compression until $\varepsilon_{vol} = 20\ \%$. Sketch the *s':t:e* diagram.

10.3. From the same clay of the previous exercise, a specimen was obtained and isotropically consolidated under $\sigma'_c = 1000$ kPa. Then, it was allowed to swell back to $\sigma'_c = 50$ kPa and was finally sheared in drained triaxial compression. Sketch the *s':t:e* diagram.

10.4. Estimate ϕ' for Rio's clay through Kenney (1959) and Mayne (1980) correlations. Given: $LL = 120\ \%$, $PL = 40\%$.

10.5. Explain the reasons which prevent a highly overconsolidated clay to reach critical state in a drained triaxial test. Sketch the typical behaviour for this clay in the *s':t:e* diagram.

Chapter 11

Undrained behaviour of clays

Introduction

This chapter deals with the behaviour of clays in *CIU* testing. They are used to represent clays in impeded drainage conditions, such as during rapid construction of embankments on clays, at the final stage of a excavation or a foundation on clay, and in the clay core of a dam at the end of construction.

Saturated soils will be considered, but some aspects relative to unsaturated clays will be pointed out.

CIU test phases

Figure 11.1 presents the phases of a *CIU* triaxial test, as studied in chapter 8. It starts with the application of a confining stress σ_c, leading to an increase of pore pressure Δu within the specimen. Consolidation is, then, allowed to take place until full Δu dissipation. This phase lasts typically between 24 and 48 hours. At the end, the specimen has changed its volume and Δu is zero.

Drainage valves are closed and the undrained shear phase starts. If the soil is saturated there is no volume change. Pore pressures are measured throughout this phase. The deviator stress is progressively increased at a controlled rate, so that pore pressure distribution within the specimen is uniform. The duration of shear loading is usually between 8 and 36 hours, during which axial load, deformation and pore pressures are recorded.

Normally consolidated clays

The behaviour of *NC* clays can be understood from typical data on the Rio de Janeiro clay. Figure 11.2 presents the results of a *CIU* test under the confining stress $\sigma'_c = 150$ kPa. Curves of $t = (\sigma_1 - \sigma_3)/2$ and Δu versus ε_1 are shown.

The results indicate the maximum shear stress t_{max} was achieved for rela-

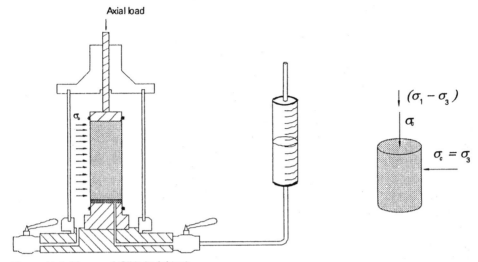

Figure 11.1. Phases of *CIU* triaxial test

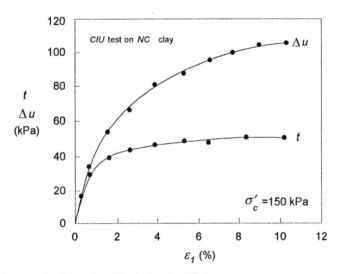

Figure 11.2. Results of a *CIU* test on Rio do Janeiro *NC* clay

tively small axial strains, in the order of 2%, where *failure* took place. There-after, the strength remained constant until the end of the test. Pore pressures increased steadily and stabilized only at large strains, about 10%. At this point, the test was stopped because specimen shape was too distorted due to excessive deformation and measurements of strains became meaningless.

Overconsolidated clays

Overconsolidated saturated clays will be studied through data published by several researchers. One of the most significant works has been carried out at the Imperial College by Henkel (1960) and Bishop and Henkel (1962). Their data were analysed by many others (e.g., Atkinson and Bransby, 1978, Lambe and Whitman, 1979, Ladd, 1971).

Figure 11.3 presents typical results of a heavily *OC* clay, (figures 11.3c and 11.3d), and compares them with the *NC* clay (figures 11.3a and 11.3b).

The *OC* clay presents positive pore pressures in the beginning of the test, but they gradually decrease and become negative (figure 11.3d). At very large strains pore pressures reach a steady negative value and stop changing. Strength *t* (figure 11.3c) also reaches a maximum value with large deformation.

Critical state

Both *NC* and *OC* saturated clay specimens tested in undrained conditions tend, at large strains, to a steady state, in which the shear strength (q or t), and the pore pressure u and (as will be shown later) the mean effective stresses p' or s' do not change any more. This steady state is the *critical state*, mathematically described by:

$$\frac{\partial q}{\partial \varepsilon_1} = \frac{\partial p'}{\partial \varepsilon_1} = \frac{\partial u}{\partial \varepsilon_1} = 0 \qquad (11.1)$$

or, in the MIT $s':t:e$ plot by:

$$\frac{\partial t}{\partial \varepsilon_1} = \frac{\partial s'}{\partial \varepsilon_1} = \frac{\partial u}{\partial \varepsilon_1} = 0 \qquad (11.2)$$

The corresponding critical state friction angle is ϕ'_{cr}.

Stress paths in *CIU* testing

The total and effective stress paths in the *CIU* tests are obtained through equations 4.14 and 4.17. *ESP's* do not coincide with the *TSP's* because pore pressures are no longer zero.

Pore pressures during shear control the *ESP* pattern. In *NC* clays (figure 11.4a), Δu is positive and the *ESP* goes to the left side of the of the TSP. On a heavily *OC* clay, Δu can be positive in the beginning, but may turn negative in later stages of a test (figure 11.3d) and the *ESP* will plot on the right hand side of the *TSP* (figure 11.4b).

Figure 11.4 also shows a practical way for sketching an *ESP* from the *TSP* and known Δu data. First plot the *TSP*, for a given test condition. The *ESP* in then plotted point by point. As an example, if one wishes to plot point *A* on the

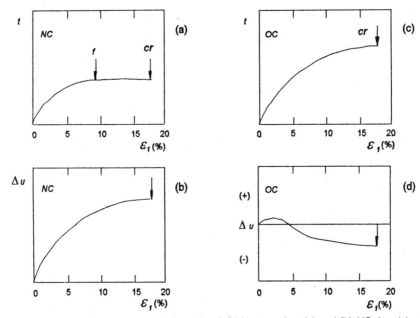

Figure 11.3. Comparison between typical results of *CIU* test on clay: (a) and (b) *NC* clay; (c) and (d) *OC* clay

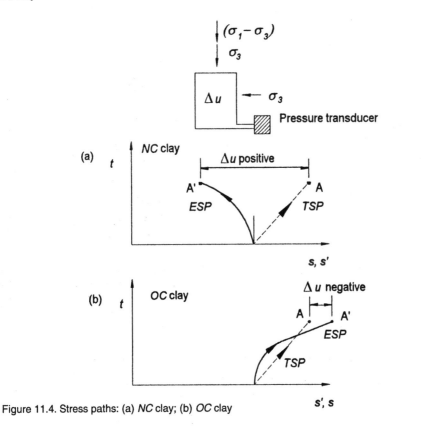

Figure 11.4. Stress paths: (a) *NC* clay; (b) *OC* clay

TSP, its equivalent *A'* on the *ESP* is shifted horizontally by an amount equal to pore pressure Δu. If Δu is positive the shift is to the left, if negative to the right.

Alternatively, one can use a spreadsheet program in microcomputer to calculate the coordinates *s':t* and plot the *ESP.*

Influence of dilation and contraction on pore pressures

The change in pore pressure Δu may be negative or positive depending on the tendency of the soil to *dilate* or to *contract*, as shown in figure 11.5.

A heavily *OC* saturated clay under a drained *CID* test exhibits dilation during shearing (see chapter 10). If the same clay is now sheared in undrained condition, soil particles will tend to dilate (figure 11.5a). But drainage is prevented. The pore water will be in tension in an effort to hold soil particles in their initial positions. Pore pressure will drop.

The opposite occurs in an *NC* clay, which decreases in volume during shearing in a drained *CID* test. If it is now tested in undrained conditions, particles will tend to compress. The pore water will oppose to this movement and pore pressure will increase.

Both cases are summarized in table 11.1.

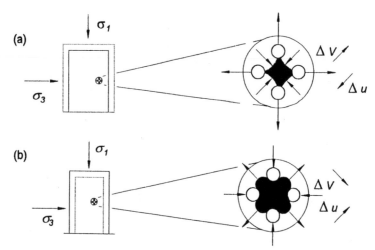

Figure 11.5. Pore pressures during shear in *CIU* tests in saturated soils: (a) tendency of dilation, leading to a decrease in Δu; (b) tendency of contraction, leading to an increase in Δu

Table 11.1. Influence of volume change tendency on pore pressure behaviour during shear

Volume change tendecy at undrained shear	Δu
Dilation	Decreases
Compression	Increases

Methods for pore pressure prediction

The element of saturated clay in figure 11.6 is subjected to the stress increments $\Delta\sigma_1$, $\Delta\sigma_2$ and $\Delta\sigma_3$ in undrained conditions. As a consequence, pore pressure varies of a amount Δu within the element.

Attempts to mathematically relate Δu with $\Delta\sigma_1$, $\Delta\sigma_2$ and $\Delta\sigma_3$ will be described.

The elastic method

If the soil is perfectly elastic and saturated, and pore water pressure is incompressible, there is no volume change during undrained shear. Entering this condition (i.e., $\varepsilon_{vol} = 0$) into equation 2.22 in terms of effective stresses, it turns out that the total mean effective stress p' is not allowed to change:

p' = constant

Any change in the mean total stress p is compensated by a change in the pore pressure u in order to keep p' constant. Therefore:

$$\Delta u = \Delta\sigma'_{oct} = \frac{\Delta\sigma_1 + \Delta\sigma_2 + \Delta\sigma_3}{3} = \Delta p \qquad (11.3)$$

The validity of this equation has been verified in practice in soft clays. Hoeg et al. (1969) observed a good agreement between Δu predicted and measured at the initial stages of loading, when soil can be regarded to be close to an elastic condition. Other researchers (Leroueil et al., 1978a, 1978b, 1985) consider that pore pressure dissipation occurring simultaneously with the initial stages of loading will lead to an error if equation 11.3 is used. In fact, field measurements in the foundations of embankments on soft soils at early stages of construction indicate that $\Delta u < \Delta\sigma_{oct}$.

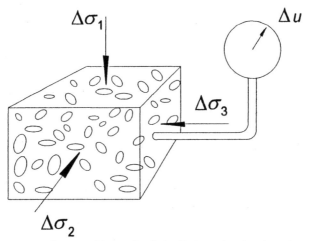

Figure 11.6. Pore pressure increment in a saturated soil element under stress increments $\Delta\sigma_1$, $\Delta\sigma_2$ and $\Delta\sigma_3$

Terzaghi's assumption

Terzaghi's assumption used in the consolidation theory was derived from equation 11.3 for one dimensional loading, leading to:

$$\Delta u = \Delta \sigma_1 \tag{11.4}$$

The use of this equation in the early stages of loading is not supported by field measurements (Hoeg et al., Leroueil et al. (*op cit*)). Observations have shown that in the plastic domain, when soil is yielding, the use of equation 11.4 leads to good agreement with field measurements (Ortigao et al., 1983).

Skempton's method

Recognizing the limitations of the elastic method for pore pressure predictions during triaxial testing, Skempton (1954) proposed the following empirical equation:

$$\Delta u = B \left[\Delta \sigma_3 + A \left(\Delta \sigma_1 - \Delta \sigma_3 \right) \right] \tag{11.5}$$

where: A and B are empirical pore pressure parameters obtained experimentally. If soil is fully saturated, then $B = 1$, and the above equation is reduced to:

$$\Delta u = \Delta \sigma_3 + A \left(\Delta \sigma_1 - \Delta \sigma_3 \right) \tag{11.6}$$

Skempton's method was devised for axisymmetrical conditions of triaxial testing, in which the effect of the intermediate principal stress increment $\Delta \sigma_2$ is not taken into account.

Exercise 11.1

Obtain Skempton pore pressure parameters for failure and critical state conditions for the *CIU* test on the Rio de Janeiro clay whose results are presented in figure 11.2.

Solution

In a *CIU* compression test in a saturated clay: $\Delta \sigma_3 = 0$ and $B = 1$. Equation 11.6 simplifies to:

$$\Delta u = A \, \Delta \sigma_1$$

$$\therefore A = \Delta u / \Delta \sigma_1 \tag{11.7}$$

As σ_3 = constant, $\Delta \sigma_1$ is given by $\Delta \sigma_1 = (\sigma_1 - \sigma_3) = 2t$. The following table summarizes data from figure 11.2 and A values corresponding to failure and critical state.

Condition	ε_1	σ_3	t	Δu	$\Delta \sigma_1$	A
	%	kPa	kPa	kPa	kPa	kPa
Failure	2	150	40	60	80	0.75
Critical state	10.5	150	45	105	90	1.17

Henkel's method

Recognizing limitations in equation 11.5, Henkel (1960) proposed an analogous equation, but including the effect of $\Delta\sigma_2$ through the octahedral stress increments:

$$\Delta u = \beta \left[\Delta\sigma_{oct} + 3\,\alpha\,\Delta\tau_{oct} \right] \tag{11.8}$$

where: α and β are empirical pore pressure parameters. In saturated clays β is equal to 1.

Replacing the values of p and q from equation 4.16 in equation 11.5 it becomes:

$$\Delta u = \beta \left[\Delta p + \alpha \sqrt{2}\,\Delta q \right] \tag{11.9}$$

Exercise 11.2

Obtain a relationship between Henkel and Skempton's pore pressure parameters α and A for a *CIU* compression test in saturated clay.

Solution
Considering axi-symmetrical and saturated conditions and constant confining stress during compression, then: $\Delta\sigma_3 = \Delta\sigma_2 = 0$, $B = \beta = 1$, $\Delta p = \Delta\sigma_1/3$, $\Delta q = \Delta\sigma_1$. Using equations 11.7 and 11.9:

$$\Delta u = \Delta\sigma_3 + A\left(\Delta\sigma_1 - \Delta\sigma_3\right) = A\,\Delta\sigma_1$$

$$\Delta u = \Delta p + \alpha \sqrt{2}\,\Delta q = \frac{\Delta\sigma_1}{3} + \alpha \sqrt{2}\,\Delta\sigma_1$$

Eliminating Δu:

$$A\,\Delta\sigma_1 = \frac{\Delta\sigma_1}{3} + \alpha \sqrt{2}\,\Delta\sigma_1$$

Simplifying and rewriting to get α:

$$\alpha = \frac{3\,A - 1}{3\sqrt{2}} \tag{11.10}$$

Exercise 11.3

Obtain Henkel's pore pressure parameters from *CIU* test data of Rio de Janeiro clay in figure 11.2.

Solution
From test conditions: $\beta = 1$, $\Delta\sigma_3 = 0$, $p_0 = 150$ kPa, $q_0 = 0$. Additional input data for the application of equation 11.9 are presented in the following table:

Condition	ε_1	σ_3	Δq	Δu	$\Delta\sigma_1$	Δp	α
	%	kPa	kPa	kPa	kPa	kPa	
Failure	2	150	80	60	80	27	0.29
Critical state	10.5	150	90	105	90	30	0.59

Alternatively, α can be obtained through equation 11.10 through A values from previous exercise 11.1.

Values of pore pressure parameters

Table 11.2 summarizes values of pore pressure parameter A_f at failure from triaxial compression testing of different clays given by Skempton (1954).

Figure 11.7a gives an indication of the *ESP'* direction, as a function of the A_f value.

Elastic and saturated materials present, as discussed before, $p' =$ constant and $\Delta u = \Delta p$. Applying these conditions to Skempton and Henkel equations they give:

$$A = 1/3$$
$$\alpha = 0$$

As a consequence, the *ESP* for elastic materials ($p' =$ constant) in the Cambridge $p':q$ plot (figure 11.7b) is a vertical line.

The effect of the degree of saturation S on the pore pressure parameter B is shown in figure 11.8. A remarkable drop in B is observed even for a small decrease in the degree of saturation from 100 to 95%. Soils presenting $S < 95\%$, generate little pore pressure, having B lower than 0.4.

Table 11.2. A_f values for different clays (Skempton, 1954)

Soil type	A_f
Soft sensitive clay	0.75 - 1.5
Soft *NC* clay	0.50 - 1.0
Compacted clay	-0.25 - 0.5
Stiff *OC* clay	-0.50 - 0.0

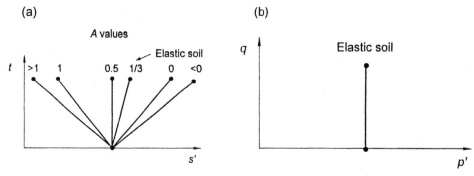

Figure 11.7. Stress paths: (a) Direction of *ESP*'s as a function of the pore pressure parameter A; (b) *ESP* in the Cambridge plot for an ideal elastic soil

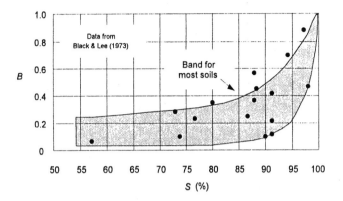

Figure 11.8. Pore pressure parameter *B* for unsaturated soils (Black and Lee, 1973)

Behaviour of *NC* clays in the *s':t:e* diagram

Figure 11.9 presents the *s':t:e* diagram for a typical *NC* clay in *CIU* testing. Figure 11.9a presents a plot of Δu and t versus ε_1, where failure (point *C*) and critical state points (*D*) are indicated. The stress paths are shown in figure 11.9b. The *TSP is AB* and the *ESP, ACD*. The maximum deviator stress t_{max} is reached at *C*. Pore pressures are still rising at *C* and continue changing until they reach the critical state at *D*, where Δu becomes constant. The failure and the critical state lines K_f and K_{cr} are not coincident.

In the *s':e* diagram (figure 11.9c) the *NC* clay specimen starts at the *ICL* (point *A'*). As shear is undrained and the clay is saturated, there is neither volume, nor void ratio change. The path is, then, a horizontal line from *A'* until it reaches the *CSL* at *D'*. The same occurs in the log *s':e* diagram (figure 11.9d), the clay follows the *A"D* path.

Behaviour of *OC* clay in the *s':t:e* diagram

Figure 11.10 presents the typical behaviour of an *OC* specimen in a *CIU* test. The specimen was initially *NC* at point *O'* on the *ICL*. Then it was allowed to swell towards point *A'* in the dry side of the *CSL*, becoming heavily overconsolidated. Then, it was sheared in undrained conditions. The resulting *ESP* is shown in figure 11.10a, and will be analysed by means of figure 11.11.

Figure 11.11 presents more detailed set graphs of the shear test. Stress-strain and pore pressure results versus strain are shown in figures 11.11a and 11.11b. Points B_1 and B_2 correspond to failure (t_{max}). At this stage, the triaxial test had to be interrupted due to excessive sample distortion. It is assumed, however, that, if strain continued, the *OC* clay would eventually reach the critical state at *C*.

The *ESP* is plotted in figure 11.11c. Point *B* lies on the K_f line, above the

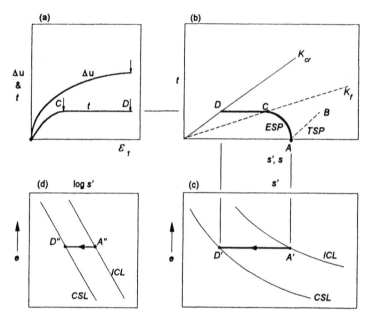

Figure 11.9. *s':t:e* diagram for *CIU* test in *NC* clay

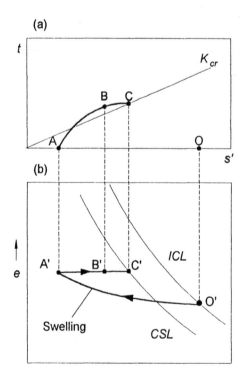

Figure 11.10. *s':t:e* diagram for *CIU* test in *OC* clay

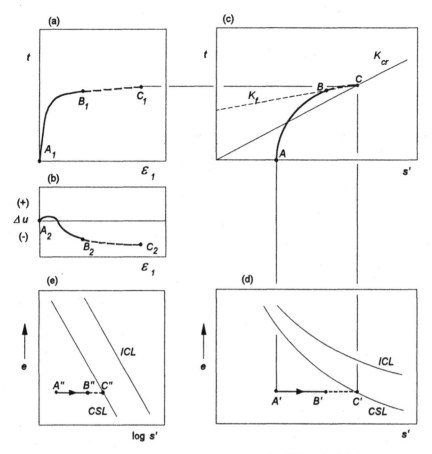

Figure 11.11. Stress-strain, pore pressure and $s':t:e$ diagram for *CIU* test in *OC* clay

K_{cr} line, in the meta-stable region. Since pore pressures are negative, the *ESP* tends to the right side and would eventually reach critical state at *C*.

Figures 11.11d and 11.11e show the $s':e$ and $\log s':e$ diagrams in which the paths during shear with no volume change are horizontal.

Behaviour of clays with the same void ratio but different *OCR*'s

Consider a set of specimens of the same clay and the same void ratio e_0 before shear, but consolidated at different confining pressures and *OCR's*. They are brought to failure in undrained shear and follow a horizontal path in the $s':e$ diagram towards the critical state at point *C'* of the *CSL* (figure 11.12b).

Figure 11.12a presents the *ESP*'s which tend to reach the same point *C* on the K_{cr} line, *OC* specimens with high *OCR* start from the dry side, reach the meta-stable region, touch the K_f line and tend to reach point *C*. The NC clay starts from the wet side, at the right of the plot, and shows a leftward move towards

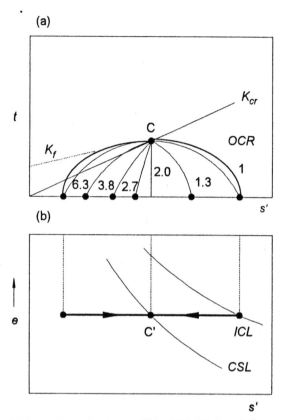

Figure 11.12. *s':t:e* diagram for several clay samples with the same pre-shear void ratio *e*, but different *OCR's*. (a) *t:s'* diagram; (b) *e:s'* diagram

point *C*. This *ESP* for the *NC* specimen is the rightmost path for this clay at this void ratio, therefore it is a *boundary limit*.

State boundary surface

The *ESP's* from figure 11.12 were replotted in figure 11.13. The *ESP's* are limited: on the upper left side, by the K_f line; on the right one, by the *ESP* of the *NC* clay. These limits constitute what is called *State Boundary Surface, SBS*.

A *SBS* is linked to a particular e_0 value, as shown in figure 11.14. If the void ratio varies during a test, the *SBS* expands or contracts.

Drained and undrained tests and the critical state envelope

The critical state model assumes the existence of a single strenght envelope, or K_{cr} line, for each soil type, regardless of the testing condition. Therefore, if an undrained test and a drained one are carried out starting at the same point *A*

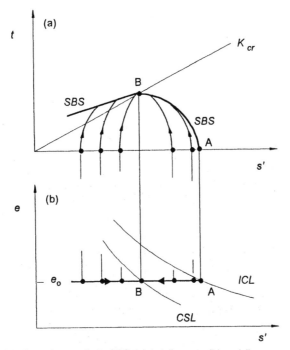

Figure 11.13. The state boundary surface *SBS*. (a) *t:s'* diagram; (b) *e:s'* diagram

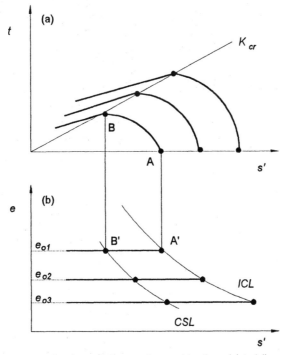

Figure 11.14. *SBS* family as a function of the pre-shear void ratio *e*. (a) *t:s'* diagram; (b) *e:s'* diagram

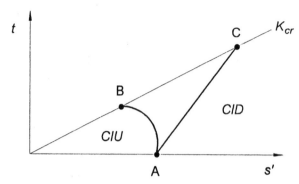

Figure 11.15. *ESP*s for *CIU* and *CID* triaxial testing on identical specimens of *NC* clay at the same consolidation pressure.

(figure 11.15), the critical state is reached at points C and B on the same K_{cr} line.

Soft clay behaviour in the light of critical state model

The theoretical framework of critical state is a simple and useful way of interpreting the behaviour of clays. It incorporates the main features of clay behaviour in a simple model. But how do real clays fit into this model?

To answer this question we will use results from over a hundred triaxial tests on the Rio de Janeiro clay, which is a typical soft deposit. We will plot them in convenient ways to compare them to the critical state model.

Figure 11.16 summarizes the results of many undrained triaxial tests. The upper plot (figure 11.16a) shows results of *NC* specimens from 2 to 6 m depth. They were consolidated in the laboratory under effective stresses greater than 100 kPa, which are several times higher than the in situ overconsolidation stress. In fact, in situ σ'_{vm} for this clay ranges from 15 kPa at the top to a maximum of 40 kPa at the middle of the clay (figure 6.18). The *ESP*'s are typical of *NC* clays and turn to the left. The K_f line is plotted. Strenght parameters at maximum deviator stresses are $c'=0$ and $\phi'=25°$.

Figures 11.16b and 11.16c show a zoomed view of the $s':t$ plot, amplifying the small stress range. All specimens here were consolidated in the laboratory under low effective stresses. They still *remember* the in situ σ'_{vm}, and are overconsolidated. Accordingly, the *ESP*'s show a leftward turn and reasonably fit the theoretical framework discussed before. Overconsolidation leads to an effective cohesion in the K_f line, shown in the lower plot.

Another *CIU* testing series is presented in figure 11.17. Overconsolidation was simulated on specimens in the laboratory by consolidating under a maximum confining pressure σ'_{vm} of 300 to 500 kPa and later allowing them to swell under a reduced consolidation pressure σ'_v. The resulting *OCR* varied from 1 to

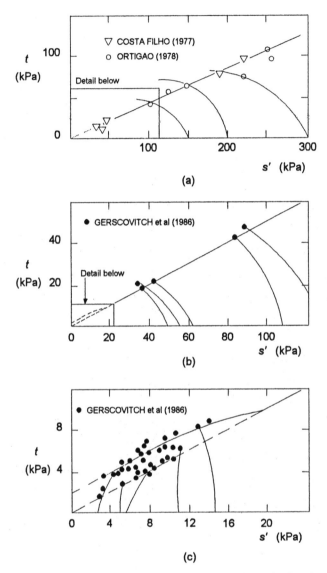

Figure 11.16. Results of triaxial tests on Rio de Janeiro clay at a wide range of confining stresses

8. The $s':e$ plot was normalized against σ'_{vm} used in the laboratory. The K_f and the K_{cr} lines are drawn.

The existence of the *CSL* was evaluated from several drained and undrained tests plotted in figure 11.18. It includes compression and extension tests, which will be studied later in chapter 13. The stress range of p' in the figure is very large. As a result, both the *CSL* and the *ICL* curved, not straight lines as assumed in the model. However, if the stress range is shortened, the *CSL* and the *ICL* could be assumed to be straight lines, and the critical state model is applicable.

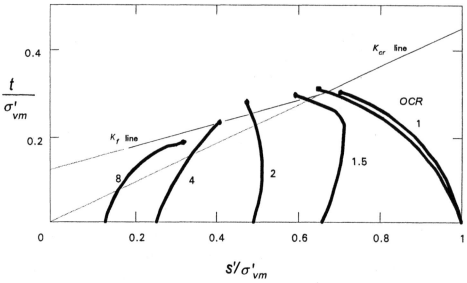

Figure 11.17. Normalized *ESP's* for *CIU* tests on Rio de Janeiro clay

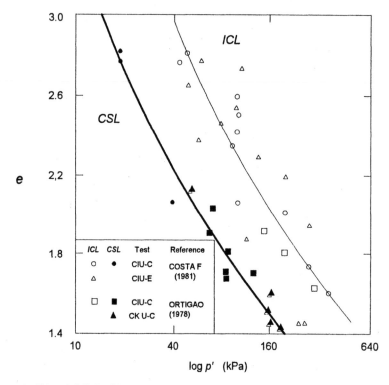

ICL	CSL	Test	Reference
o	•	CIU-C	COSTA F
△	▲	CIU-E	(1981)
□	■	CIU-C	ORTIGAO
	▲	CK U-C	(1978)

Figure 11.18. *ICL* and *CSL* for Rio de Janeiro clay (Almeida, 1982)

Exercise 11.4

A *NC* clay specimen was submitted to a *CIU* triaxial compression test and presented the *ESP AC* (figure 11.19a). Given: the *ICL* and the *CSL* in figure 11.19b. Plot the paths in the *s':e* diagram and (b) calculate the pore pressure parameters *A* and α at the critical state.

Solution

(a) The *s':e* plot:

Point *A* of the *ESP* corresponds to *A'* on the *ICL* in the *s':e* plot. As there is no volume change, the path *A'C'* is horizontal and reaches critical state at *C'* on the *CSL*.

(b) Pore pressure parameters *A* and α :

As $\Delta\sigma_3 = 0$, the value of *A* can be obtained from equation 11.7. Values of $\Delta\sigma_1$ and Δu can be obtained graphically on the *MIT* plot, as shown in the figure. The results are summarized below:

Sample	ESP	TSP	Δu	$\Delta\sigma_1$	A
NC	AC	AE	EC = 230 kPa	AG = 190 kPa	1.21

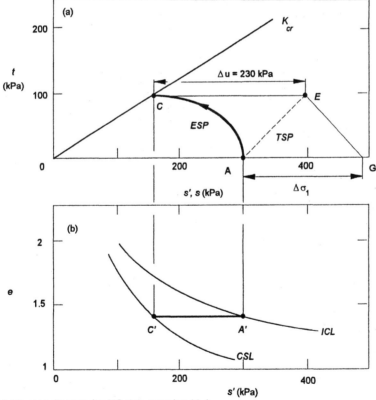

Figure 11.19. *s':t:e* diagram for *NC* clay, exercise 11.4

Exercise 11.5
Repeat previous exercise for the *OC* clay shown in figure 11.20. The initial void ratio before shear is the same as the *NC* of the previous exercise.

Solution
(a) Path in the $s':e$ plot:

The pre-shear e is obtained in the previous problem. The *OC* clay will reach the critical state at the same point C' as in the previous exercise. The path in the $s':e$ plot (figure 11.20b) is horizontal, points B' and D' are determined.
(b) The pore pressure parameters A and α are:

Sample	ESP	TSP	Δu	$\Delta\sigma_1$	A
OC	BDC	BF	FC = 24 kPa	BH = 190 kPa	-0.13

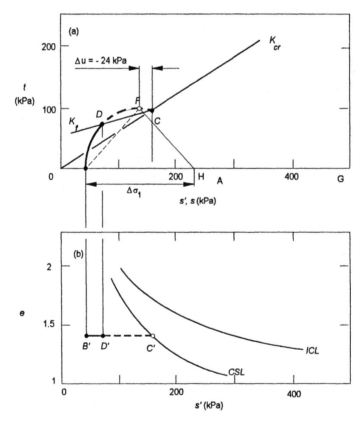

Figure 11.20. $s':t:e$ diagram for *OC* clay, exercise 11.5. (a) $t:s'$ diagram; (b) $e:s'$ diagram

Exercise 11.6

Consider 4 clay specimens of the same clay with the same void ratio at the end of consolidation. Two specimnes are *NC* and lie at point *A* (figure 11.21a) on the *ICL*, before shear. The others are *OC* and lie at *B* on the dry side before shear. One *NC* and one *OC* specimen is subjected to drained shear, the others to undrained shear. Given: the *CSL, ICL, K_f* and *K_{cr}* lines. Plot the *s':t:e* diagram.

Solution

The *NC* specimen subjected to the undrained test, has an horizontal path *A'C'* in the *s':e* plot and reaches the critical state at *C'* on the *CSL*. From *C'*, point *C* is obtained on the *K_{cr}* line.

The drained test on the *NC* specimen presents an *ESP* with known 1:1 slope reaching *F* on the *K_{cr}* line. Then, *F'* is obtained in the lower plot on the *ISL*. The path *A'F'* is drawn.

The undrained shear on the *OC* specimen, starts its at *B'* on the dry side in the lower plot, and follows an horizontal path towards *C'* on the *CSL*. The *ESP* touches the *K_f* line and continues towards point *C* on the *K_{cr}* line.

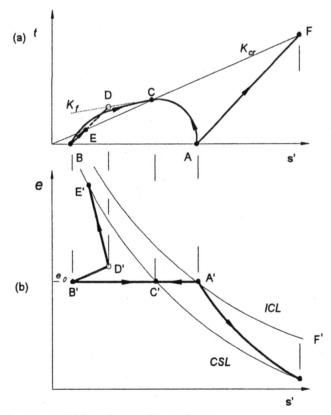

Figure 11.21. Exercise 11.6. (a) *t:s'* diagram; (b) *e:s'* diagram

Finally, the drained test on the *OC* specimen follows an *ESP*, with a known 1:1 slope until reaches point *D* on the K_f line, in the meta-stable region. Then it returns to the K_{cr} line, reaching the critical state at *E*. The path in the *s':e* diagram is *B'D'E'*, where point *D'* was arbitrarily chosen, and point *E'* is on the *CSL*.

Proposed exercises

11.1. Define critical state in undrained conditions and give its mathematical expressions.

11.2. Can the critical state parameters C_c, C_s, ϕ', e_{cs} and G be obtained from only one *CIU* test? Sketch the *ESP* of this test.

11.3. Sketch the *s':t:e* diagrams for *CIU* tests on a *NC* clay. Repeat for an *OC* clay which starts shearing from the dry side.

11.4. For the test data presented in figure 11.2, plot the *TSP* and the *ESP*, obtain ϕ' and the pore pressure parameters A and α. Are these parameters referred to the failure, or critical state condition? Obtain the undrained modulus E_u and υ_u for a stress level of 50%.

11.5. Show that for an elastic saturated material: $\Delta u = \Delta\sigma_{oct}$ and Skempton's parameter A = 1/3. Sketch its *ESP* in undrained shear either in the MIT and the Cambridge plot. What is the Henkel's a value in this case?

11.6. A *CIU* triaxial compression test on a clay sample at the confining stress of 330 kPa gave the results tabulated below.

$(\sigma_1 - \sigma_3)$ kPa	ε_1 (%)	Δu kPa
0	0	0
30	0.06	15
60	0.15	32
90	0.30	49
120	0.53	73
150	0.90	105
180	1.68	144
210	4.40	187
240	15.5	238
235	20.0	240

Plot the *s':t:e* diagram and obtain the parameters: ϕ' and the pore pressure parameters A for failure and critical state conditions. Obtain the shear modulus G in the initial part of the stress-strain curve.

Chapter 12

The $\phi_u = 0$ method and *UU* tests

Introduction

This chapter presents a simplified method of dealing with clays through total stresses, known as the $\phi_u = 0$ method, in which *UU* undrained tests are used.

The $\phi_u = 0$ method

The theoretical basis of the $\phi_u = 0$ method was established by Skempton (1948). It consists of a very useful simplified approach to obtain the strength of low permeability and saturated soils subjected to rapid loading.

Consider point P in a clay layer (figure 12.1a), subjected to the indicated loading which will be applied quickly.

The rigorous approach for analysing the shear strength at point P was studied in the previous chapter, through effective stresses. The in situ state of stress at P is represented by point A of figure 12.1b. Assuming that the state of stress at P is known before and after loading, the paths are plotted. The *TSP* is AB and the *ESP* is AC, which reaches failure at C. The value of pore pressure change at failure Δu is CB. The shear strength of point P is given by t_f.

The shear strength in terms of effective stresses is given by the Mohr-Coulomb's equation $\tau_{ff} = c' + [\sigma_{ff} - (u_0 + \Delta u)] \tan \phi'$ (or the transformed equation: $t_f = a' + (s - u) \tan \alpha'$), in which the necessary soil parameters are c', ϕ', and Δu should also be given. The value of σ_{ff} is evaluated from the loading geometry and unit weights of soils, u_0 from the groundwater regime, assumed to be known. The effective strength parameters c' and ϕ' can be obtained from *CIU* or *CID* tests. However, experience demonstrates (e.g., Bishop and Bjerrum, 1960) that the great uncertainty lies on the in situ excess pore pressure Δu during loading. Theoretical predictions of Δu using pore pressure equations from the last chapter can only give a rough estimate, the correct value has to be measured in situ. Therefore, the evaluation of the shear strength to be used in design through effective stresses is difficult.

Skempton, then, devised a simple solution for practical applications, proposing a fictitious treatment in terms of total stresses. This technique is shown in figure 12.1c. He replaced the *ESP* and and the effective strength envelope by an a fictitious horizontal envelope, where the name $\phi_u = 0$ comes from, intercepting point *B* of the *TSP*. This envelope is defined by one parameter only: the intercept at the ordinate axis, c_u, known as the *undrained shear strength*. Consequently, the strength at point *P* is $t_f = c_u$.

The differences between the methods to determine strength for undrained loading are summarized in table 12.1.

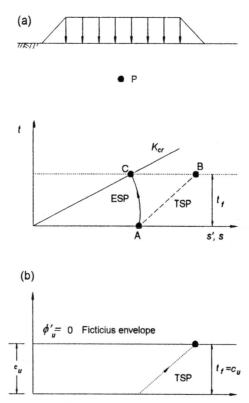

Figure 12.1. Shear strength at point *P* in the foundation of an embankment on soft clay: (a) effective stress analysis; (b) total stress analysis with ϕ_u envelope

Table 12.1. Differences between the effective stress and the total stress $\phi_u = 0$ method

Method	Strength at the failure plane	Necessary parameters
Effective stresses	$\tau_{ff} = c' + \left[\sigma_{ff} - \left(u_0 + \Delta u\right)\right]\tan\phi'$	u, c', ϕ
Total stresses $\phi_u = 0$	$t_f = c_u$	c_u

c_u from triaxial tests

The undrained shear strength is given by *UU* triaxial compression tests, provided the test is run at the same in situ void ratio e_0. Consider point *P* (figure 12.2a) at the in situ void ratio e_0. An undisturbed sample is obtained and tested in the laboratory at the same e_0 (figure 12.2b). The resulting stress-strain curve is shown in figure 12.2c. The height of the $\phi_u = 0$ envelope is given by $t_f = t_{max} = c_u$ (figure 12.3).

c_u is given by:

$$c_u = t_f = \frac{(\sigma_1 - \sigma_3)_f}{2} \qquad (12.1)$$

(a)

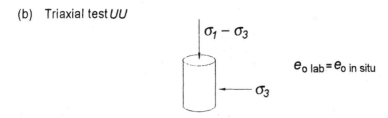

P   in situ e_0

(b)   Triaxial test *UU*

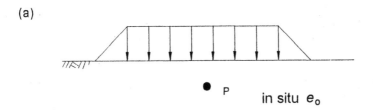

$\sigma_1 - \sigma_3$

$e_{0\ lab} = e_{0\ in\ situ}$

σ_3

(c)

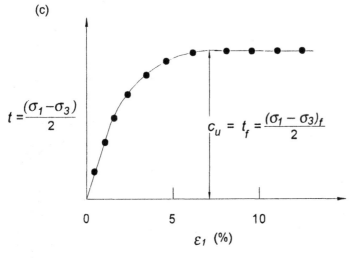

$t = \frac{(\sigma_1 - \sigma_3)}{2}$

$c_u = t_f = \frac{(\sigma_1 - \sigma_3)_f}{2}$

0   5   10

ε_1 (%)

Figure 12.2. Shear strength in the soft foundation obtained through an *UU* triaxial test

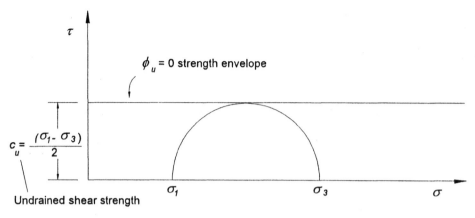

Figure 12.3. Mohr's circle and the $\phi_u = 0$ strength envelope

Exercise 12.1

The results of a *UU* triaxial test conducted at $\sigma_3 = 100$ kPa on Rio de Janeiro clay sample from 4.5 m depth are shown in the table below. Plot the stress-strain curve and obtain c_u.

ε_1	$(\sigma_1 - \sigma_3)$
%	kPa
0.2	2.5
0.5	4.8
1.0	8.5
1.5	11.0
3.0	12.5
5.5	13.5
8.0	14.0
11.0	14.0
13.0	14.0
15.0	14.0

Solution

The stress strain curve is plotted in figure 12.4. The value of $(\sigma_1 - \sigma_3)_{max}$ is 14 kPa. Therefore, through equation 12.1: $c_u = 14/2 = 7$ kPa.

The undrained strength in the laboratory and in situ

How to reproduce in the laboratory the undrained strength c_u which will be mobilized in situ during construction? The answer is provided by special *UU* test with pore pressures measurements. From the test data, we plot the $s':t:e$ diagram in figure 12.5. It is assumed that the laboratory sample is perfectly undisturbed

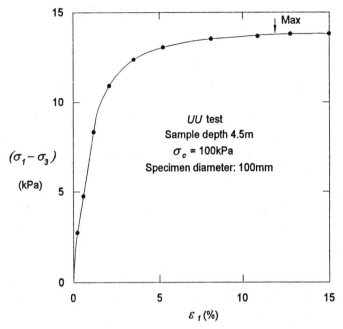

Figure 12.4. *UU* triaxial test results on Rio de Janeiro clay sample, 4.5 m depth, confining stress 100 kPa, specimen diameter 100 mm

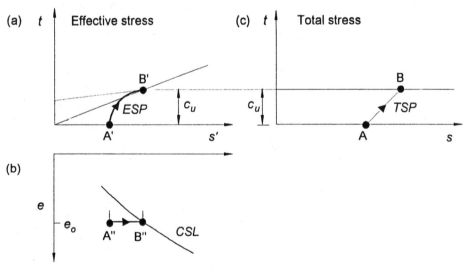

Figure 12.5. *ESP* and *TSP* in an *UU* triaxial test

and its void ratio e_0 is the same as in situ. As the specimen is sheared, it follows the path $A''B''$ in the $s':e$ diagram (figure 12.5b), reaching the *CSL* at point B''. Since there is no change in the void ratio e_0, the test exactly reproduces field conditions. The corresponding *ESP* is $A'B'$, indicated in figure 12.5a.

Therefore, the in situ value of c_u is obtained d in the laboratory; regardless of the total stress path AB adopted in the UU test (figure 12.5c) or regardless of the total confining stress (point A of the TSP), we assume that the critical state is always reached at points B'' and B', and the correct in situ value of c_u is obtained.

c_u profile

A c_u profile can be worked out from results of several UU tests carried out at different depths, as indicated in figure 12.6. Then, the geotechnical engineer makes a choice about the c_u profile to be adopted for design purposes.

Figure 12.7 presents profiles from three additional sites, two of them of *NC* clays, another of *OC* clay. The figure shows that c_u increases with depth in the *NC* clays, but in the *OC* clay it is practically does not vary with depth. This is explained in figure 12.8.

Consider the profile of figure 12.8a. The clay is *OC* between points A a B, becoming *NC* below. Samples from points A, B and C were obtained and subjected to undrained shear at the same in situ void ratio with pore pressure measurements. The resulting *ESP's* are shown in figure 12.8b. In the *OC* range the strength is given by the K_f line. Between points A' and B' (figure 12.8b) the strength does not change significantly, therefore c_u remains nearly unchanged. In contrast, in the *NC* range the strength is controlled by the K_{cr} line which is more inclined than the K_f line. Consequently, the undrained strength c_u between points B and C varies significantly.

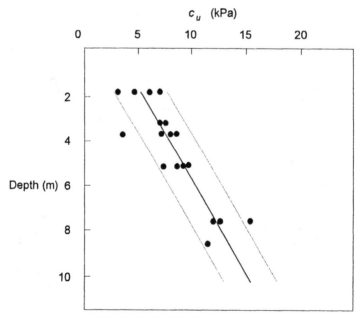

Figure 12.6. *UU* triaxial tests on Rio de Janeiro clay (Costa-Filho, 1977)

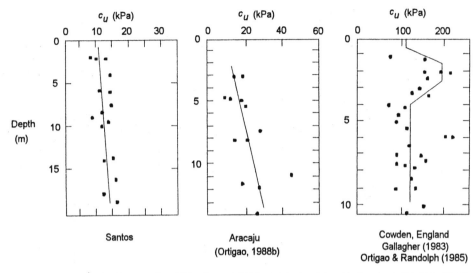

Figure 12.7. Undrained strength profiles from *UU* triaxial tests: (a) soft clay from Santos, Brazil (Teixeira, 1988); (b) soft clay from Aracaju, Brazil (Ortigao, 1988); (c) stiff *OC* clay from Cowden, England (Gallagher, 1983, Ortigao and Randolph, 1983)

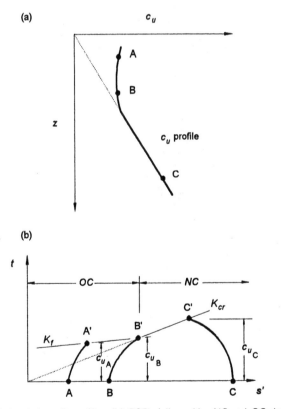

Figure 12.8. (a) Undrained strength profiles; (b) *ESP*'s followed by *NC* and *OC* clay samples

Influence of sample disturbance

c_u data in figures 12.6 and 12.7 present a considerable amount of data scatter as a result of two factors. One is the natural variability of soil deposits, the other is the disturbance due to sampling, storage, transportation and handling, which is impossible to avoid, but can be minimized.

Experience demonstrates that one important way to reduce disturbance is by taking high quality large diameter samples. c_u values are particularly influenced by the sample quality. This is demonstrated by the data in figure 12.9. The larger the sample diameter, the larger the c_u profile.

The unconfined compression *U* test

The unconfined compression, or *U* test, is a special type of triaxial test that involves axial compression of a cylindrical specimen under zero confining stress. Theoretically it should give the same results as the *UU* triaxial test.

However, since the specimen is tested without a surrounding rubber membrane, the water content may change during testing, and the strength results may be affected.

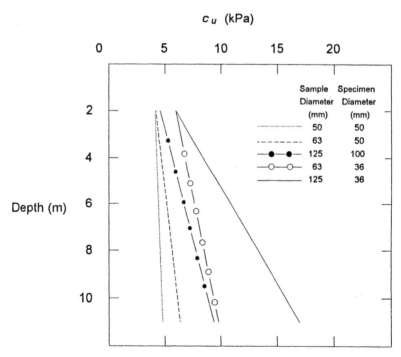

Figure 12.9. Disturbance effects in undrained strength profiles in different diameters samples and specimens (Ortigao and Almeida, 1988)

The field vane test (FV)

In situ testing is a way of avoiding disturbance effects during sampling. In particular, for the determination of the undrained strength of soft and medium clays, the field vane (FV) test has been used with success.

Early attempts to determine the undrained shear strength by means of this test occurred in Scandinavia in 1919 (Flodin and Broms, 1981), but this technique only started to be used in Europe and North America in the late forties.

The test involves inserting vertically into the ground a four blade vane (figure 12.10) and rotating it until failure is reached in 2 to 5 minutes. Efforts to standardize the vane dimensions to 130 mm height (H) and 65 mm diameter (D) have been made (Chandler, 1987) to keep a relationship $H/D = 2$. The resulting torque versus rotation curve (figure 12.11), allows the determination of the maximum torque necessary to fail the soil around the blades.

The interpretation of the test assumes that the undrained shear strength c_u is equally distributed along the cylindrical surface around the vane with $H/D = 2$, leading to:

$$c_u = 0.86 \frac{T}{\pi D^3} \tag{12.2}$$

where: T is the maximum applied torque (kNm); D is the vane diameter (m).

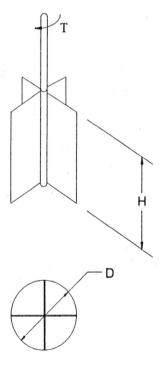

Figure 12.10. The vane

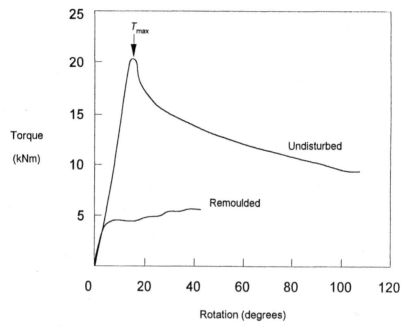

Figure 12.11. Torque-rotation curves from field vane test in undisturbed and remoulded clay

Several types of equipment are used in practice, but they differ in quality (Ortigao and Collet, 1987). Figure 12.12 presents the one of the best type, designed by Cadling and Odenstad (1950).

It involves a system of protected rods which transmit the torque to the vane. These rods are fully protected against contact with the soil, thus avoiding undesirable friction. The vane is installed recoiled within a protection shoe to avoid damage when encountering hard materials. At the required depth, the rods are released and the vane is pushed 0.5 m into the ground, and rotated. Usually tests are conducted in constant 0.5 or 1 m intervals along depth.

After testing the undisturbed clay, the vane is rotated 10 to 20 times to disturb the clay completely and the test is repeated. The *remoulded undrained strength* c_{ur} is obtained through equation 12.2, using the torque corresponding to the remoulded clay.

Results are plotted as shown in figure figure 12.13, giving profiles of both c_u and c_{ur}. Results of several boreholes in the same site can be plotted together, as indicated in figure 12.14, allowing the analysis of mean values, variability in soil properties and selection of the design strength profile.

Sensitivity

The ratio between the undisturbed and the remoulded undrained FV strength is

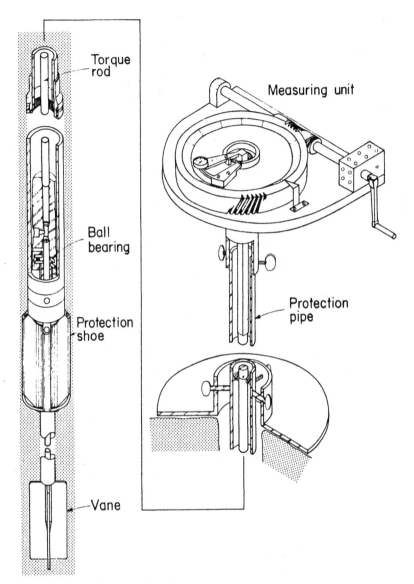

Figure 12.12. Details of field vane test equipment

defined as *sensitivity* S_t of a clay:

$$S_t = \frac{c_u}{c_{ur}} \qquad (12.3)$$

This parameter indicates the relative loss of strength due to remoulding, and the importance of the structure of the clay. Although, this can be obtained by means of laboratory tests, the FV provides a very simple and cheap way for assessing the sensitivity.

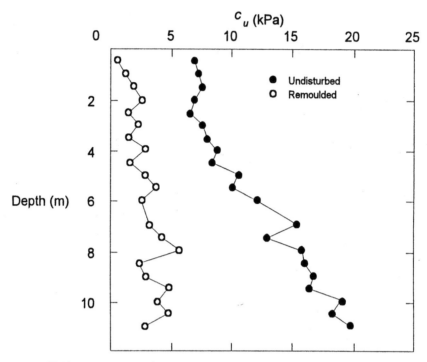

Figure 12.13. Field vane test results from a complete borehole in Rio de Janeiro clay

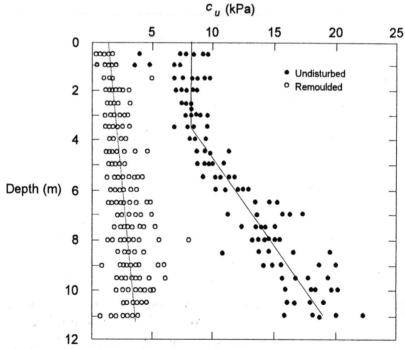

Figure 12.14. Summary of *FV* test results from several boreholes in Rio de Janeiro clay (Ortigao and Collet, 1987)

Table 12.2. Sensitivity of clays (Skempton and Northey, 1952)

Sensitivity	S_t
Low	2 - 4
Medium	4 - 8
High	8 - 16
Very high	> 16

Table 12.3. Sensitivity of some clay deposits

Site	Mean value	Range	Reference
Laing Bridge, Richmond, Canada	30	15-35	Author's experience on board the UBC site investigation unit
Cloverdale, Vancouver, Canada	17	8 to 29	Greig et al., 1987
Rio de Janeiro, Brazil	4.4	2 - 8	Ortigao & Collet (1987)
Santos, Brazil		4 - 8	Teixeira (1988)
Sergipe, Brazil	5	2 - 8	Ortigao & Sayao (1994)

Table 12.2 presents a classification of clays according to the sensitivity proposed by Skempton and Northey (1952).

Most European, North and South American clays can be regarded as low to medium sensitive clays. Table 12.3 brings some examples. Clay deposits subjected to severe leaching may develop a very high sensitivity. This is the case of several marine deposits in Scandinavia and in Canada that reached values over 100 (e.g., Leroueil et al., 1985).

Correction of FV test results

Back-analysis of failures of embankments and excavations on soft clays has led to the conclusion that the FV test results have to be corrected in most cases through an equation of the type:

$$c_{u_{corrected}} = \mu \ c_{u_{FV}} \tag{12.4}$$

The need for correction has been interpreted by Bjerrum (1973) as due to differences in strain rate, anisotropy and creep effects in the field. The determination of the correction factor μ is an empirical approach and is based on back-analyses of failures. Correction factors have been presented by Bjerrum (1973) and Azzouz et al. (1983) as a function of the plasticity index *PI*, but they present considerable scatter. Figure 12.15 present an attempt by Aas et al. (1986) to re-

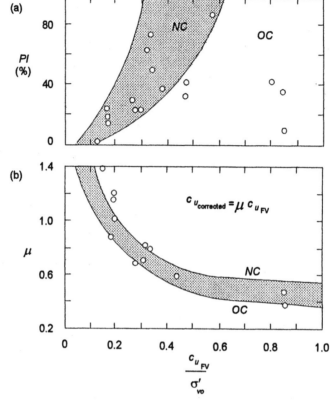

Figure 12.15. Correction of *FV* strength (Aas et al., 1986)

duce uncertainty in selecting the correction factor. Two alternatives are presented: as a function of *PI* (figure 12.15a) or, according to their suggestion, as a function of the c_u/σ'_{v0} ratio (figure 12.15b), where: c_u is the FV strength and σ'_{v0}, the in situ effective overburden stress.

FV corrections should be applied with caution. First because of the scatter in recommended correction factors, it is not easy to select values of μ. Second, because there are cases in high plasticity clays, in which corrections do not seem to be applicable and the mean FV strength profile has yielded good results (Ortigao et al., 1987 and 1988). Local experience is, therefore, the only way to solve the problem. It has been suggested, however (Ortigao and Almeida, 1988) that the c_u profile from FV should be compared to additional data from *UU* triaxial test. If there is a significant difference, correct FV data, if there isn't, don't do it.

Exercise 12.2

Apply FV correction from figure 12.15b to the mean c_u profile of figure 12.14.

Solution

The c_u/σ'_{v0} ratio is obtained at different depths. The value of σ'_{v0} was calculated assuming $\gamma = 13\ kN/m^3$. The clay was assumed OC in the first 3 m of depth, and NC below. Data are presented in the following table:

z	$c_{u_{mean}}$	σ'_{v0}	c_u/σ'_{v0}	OC/NC	μ	$c_{u_{corrected}}$
(m)	(kPa)	(kPa)				(kPa)
1	8.6	3	2.9	OC	0.4	3.4
2	8.6	6	1.4	OC	0.4	3.4
3	8.6	9	0.9	OC	0.4	3.4
4	9.4	12	0.8	NC	0.5	4.7
6	12.6	18	0.7	NC	0.5	6.3
8	14.8	24	0.6	NC	0.6	8.9
10	18.6	30	0.6	NC	0.6	8.9

Empirical determination of c_{ur}

The remoulded clay strength c_{ur} can be estimated through the empirical correlation obtained by Carrier and Beckman (1984).

$$c_{ur} = p_{atm} \left[\dfrac{0.166}{0.163 + \dfrac{37.1\, e + PL}{PI \left[4.14 + A_c^{-1} \right]}} \right]^{6.33} \tag{12.5}$$

where: p_{atm} = atmospheric pressure = 100 kPa; PI = Plasticity index (%); e = void ratio; A_c = activity of the clay; PL = plastic limit.

Figure 12.16 presents a chart for graphically solving equation 12.5.

Relationship between c_u and effective stresses

One way to evaluate the undrained strength is to relate it to effective stresses. The ratio c_u/σ'_v has been found to be very useful, since it is an unique function of the OCR. In the light of critical state theories, the following equation was deduced (e.g., Atkinson and Bransby, 1978) relating the c_u/σ'_v ratio of NC to OC samples:

$$\frac{\left(c_u/\sigma'_v \right)_{OC}}{\left(c_u/\sigma'_v \right)_{NC}} = OCR^\Lambda \tag{12.6}$$

This equation allows one to obtain the undrained strength provided we know the

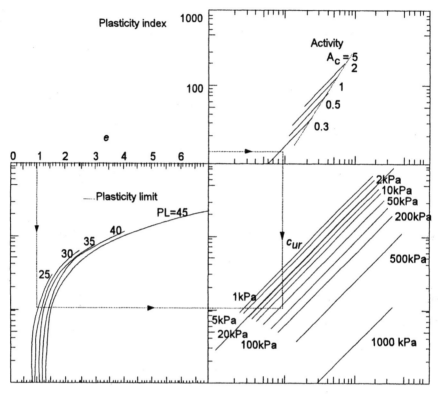

Figure 12.16. Chart for determining remoulded undrained strength from a correlation with Atterberg limits (Carrier and Beckman, 1984)

OCR and the value of parameter Λ. This forms the basis of a semi-empirical method proposed by Ladd and Foott (1974), which, presents the limitation of requiring special high quality laboratory *CU* type tests and, so, can only be employed to major projects.

Experience has shown that, in practice, the value of parameter Λ does not vary significantly, ranging between 0.7 and 0.85 (Ladd et al., 1977) and the average value of 0.8 can be adopted for most clays. A value of 0.25 can reasonably be adopted for the ratio c_u/σ'_v in *NC* clays. Then, equation 12.6 can be rewritten:

$$c_u/\sigma'_v = 0.25 \, OCR^{0.8} \tag{12.7}$$

Mesri (1975) proposed a semi-empirical equation relating c_u with the overconsolidation stress σ'_{vm}:

$$c_u = 0.22 \, \sigma'_{vm} \tag{12.8}$$

Exercise 12.3

Obtain a c_u profile for the Rio de Janeiro soft clay through equations 12.7 and 12.8. Then, compare the results with laboratory *UU* and FV data.

Solution
The table below presents the application of equations 12.7 and 12.8, where: z is the depth, σ'_{v0} and σ'_{vm} were obtained in figure 6.18. The *OCR* was calculated by equation 6.2. The c_u/σ'_v ratio was calculated by equation 12.7. c_u was obtained multiplying the ratio c_u/σ'_v by σ'_{v0}. The last column gives c_u from equation 12.8, multiplying the column of σ'_{vm} by 0.22. The results are plotted in figure 12.17.

z	σ'_{v0}	σ'_{vm}	*OCR*	Equation 12.7		Equation 12.8
				c_u/σ'_v	c_u	c_u
(m)	(kPa)	(kPa)			(kPa)	(kPa)
1	3	19	6.3	1.09	3.3	4.2
2	6	18	3.0	0.60	3.6	4.0
3	9	23	2.6	0.54	4.9	5.1
4	12	26	2.2	0.47	5.6	5.7
5	15	30	2.0	0.44	6.6	6.6
7	21	38	1.8	0.40	8.4	8.4
9	27	45	1.7	0.38	10.2	9.9
10	30	48	1.6	0.36	10.8	10.6

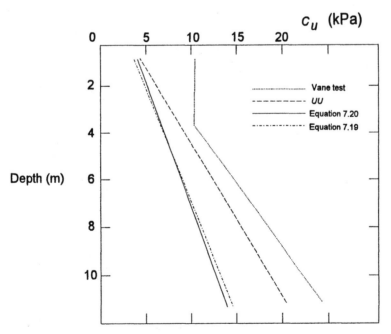

Figure 12.17. Exercise 12.3

The use of *UU* analysis and the $\phi_u = 0$ method

The *UU* type analysis employs total stresses and unconsolidated undrained type tests to analyse the stability of soil structures. It is applicable to saturated fine grained soils under rapid loading, resulting in an undrained behaviour. Therefore, the $\phi_u = 0$ method can be applicable.

A few examples are presented in figure 12.18. The mobilized strength at failure along the slip surface is τ_{ff}. Assuming $\phi_u = 0$, then: $\tau_{ff} = c_u$. Figure 12.18a presents a case of an embankment constructed on soft ground at the final stage of placement. The undrained strength of the soft foundation is assumed to be mobilized and used for the analisys. Figure 12.18b presents an embankment dam at the end of construction. The undrained strength of the clay core is employed in the stability analysis at this stage of the dam. A third example (figure 12.18c) comprises a shallow foundation on clay, just after loading. The bearing capacity is evaluated by means of the undrained strength of the clay.

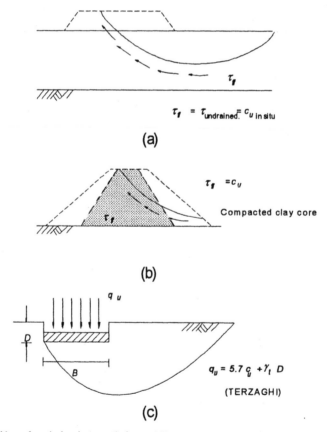

Figure 12.18. Use of undrained strength for stability analysis: (a) embankment constructed rapidly on soft clays; (b) end-of-construction stability analysis of an embankment dam with clay core; (c) shallow foundation on clay, quickly loaded (Ladd, 1971)

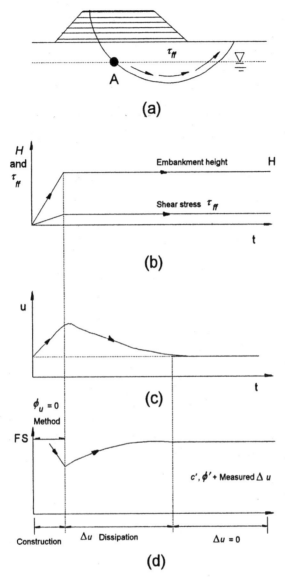

Figure 12.19. Phases of construction of an embankment on soft soil in which the $\phi_u = 0$ method can be applied (Bishop and Bjerrum, 1960)

The embankment on soft clay is presented again in figure 12.19. Consider point *A* in the clay foundation. As the embankment is placed, the mobilized stresses increase until they reach the shear strength c_u, when failure occurs. Pore pressures increase during loading and reach a maximum at the end of construction. Beyond that, they dissipate with time, until they reach equilibrium (i.e., $\Delta u = 0$).

The factor of safety *FS* decreases with loading. At the end of construction *FS* reaches the minimum value. Thereafter, it tends to increase as pore pressure dissipates, and reaches a maximum at the end of consolidation.

The $\phi_u = 0$ method should be applied for analysis of stability at the end of construction. Undrained clay behaviour is assumed. This avoids the uncertainty of pore pressure predictions. However, during and after consolidation, this method does not apply and the stability analysis should be conducted in terms of effective stresses with appropriate values of c', ϕ' and pore pressures measured in the field.

Proposed exercises

12.1. Consider a standard *UU* triaxial test in a clay specimen. Is the *ESP* known? Why? What are the results given by this test?

12.2. Present a brief description of the $\phi_u = 0$ concept applied to the end of construction analysis of an embankment on clay.

12.3. Why can *UU* triaxial tests in clay present greater data scatter than *CIU* tests?

12.4. A clay gave the following critical state parameters: $\phi' = 30°$; $C_c = 1.2$; $C_s = 0.1$; $e_{cs} = 5.3$; $e_c = 5.9$. A specimen of this material was consolidated isotropically under $\sigma'_c = 50$ kPa, and then a *UU* test was carried out. Estimate the undrained shear strength of this specimen.

12.5. Present a brief description of the field vane test.

12.6. Which tests would you recommend in the following cases? (a) An embankment on soft ground, end of construction analysis; (b) the same, 30 years after construction; (c) earth dam, end of construction analysis; (d) shallow foundation on clay; (e) shallow foundation on sand; (f) oil tank on soft clay.

Chapter 13

Applications to geotechnical engineering

Introduction

This chapter presents a few applications of the critical state model and the $s':t:e$ diagram to the analysis of practical problems in geotechnical engineering. First, a review of the most frequent stress paths, other than compression, will be presented. Then, applications to the analysis of retaining walls, embankments, excavations and piles will be reviewed.

The final topic deals with the residual state, not covered by the critical state model.

Classification of stress paths

Stress paths can be classified according to the type of loading and its direction (figure 13.1). Four main types are distinguished: axial compression, axial extension, lateral extension and lateral compression, as follows:

Axial compression

Axial compression occurs under the axis of an embankment (figure 13.1a). The vertical stresses increase and the correspondent stress path follows an upward direction to the right with a 1:1 slope. Most triaxial tests are loaded in the same way. It is convenient to apply axial compression by increasing the load on the ram, keeping the cell pressure constant. This is the reason why, in the preceding chapters, only axial compression was discussed.

Axial extension

Axial extension occurs under the axis of an excavation (figure 13.1b) in which the vertical stress decreases and the horizontal stresses are kept approximately constant. The corresponding triaxial tests are *CIU-E* or *CID-E*, in which *E* stands for Extension. The extension path follows a downward direction inclined to the left with sloping 1:1.

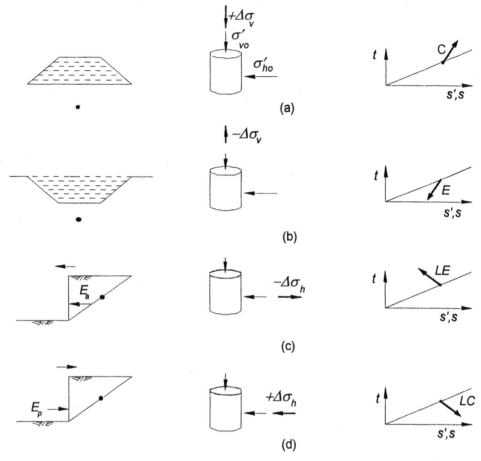

Figure 13.1. Classification of stress paths: (a) axial compression; (b) axial extension; (c) lateral extension; (d) lateral compression

Lateral extension

Lateral extension consists of decreasing the horizontal stress, keeping the vertical stress constant. It occurs in a soil element of the backfill of a retaining wall (figure 13.1c), if the wall tends to move outward, due to soil pressure. As a consequence, the horizontal stress in the soil element decreases, while vertical stresses remain unchanged. The lateral extension test suffix is *LE*, for example, *CIU-LE* or *CID-LE*.

Lateral compression

Lateral compression simulates a soil element behind a retaining wall subjected to an external loading, which moves it against the backfill. This is the case of a retaining wall receiving an external horizontal load from a foundation (figure 13.1d). The soil element of the backfill will be stressed horizontally, without significant changes in the vertical stress. The corresponding stress path is directed downwards, inclined 1:1 to the right hand side.

Laboratory lateral compression tests have the suffix *LC*, and are known as *CIU-LC* or *CID-LC*.

The *UKU* test

The *UKU* test has been used to simulate a condition of a clayey soil element in an embankment during construction and to study its pore pressure and strength behaviour (figure 13.2). It was devised to model the stress condition occurring in a embankment. The total principal stresses σ_1 and σ_3 in an embankment increase at a constant $K = \sigma_3/\sigma_1$ ratio. In the laboratory a *UU* test is conducted with constant stress ratio K and pore pressure measurements. Eventually, the specimen is brought to failure in undrained conditions by axial compression. These tests have been extensively used for the study of the behaviour of compacted residual clays used in dams (Cruz, 1967).

K_0 anisotropic consolidation

The effect of pre-shear stresses can be studied in the laboratory by consolidating specimens under the same stresses as exhibited in the ground. Therefore, a K_0 stress path shown in figure 13.3 is used. Portion AB is on the virgin K_0 line and the *ICL*. Overconsolidation can be simulated by allowing the specimen to

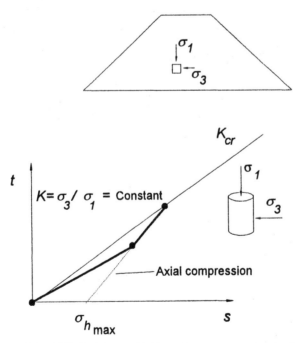

Figure 13.2. Stress path in *UKU* triaxial test with $K = \sigma_3/\sigma_1$ = constant for simulation of rapid construction of earth dam

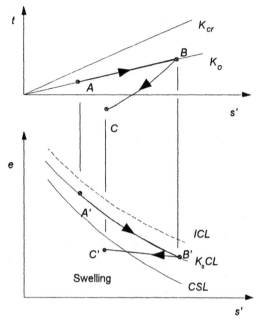

Figure 13.3. K_0 consolidation path

swell, as shown in portion *BC* in the same figure. Triaxial tests that simulate in situ stresses prior to shear are called CK_0U , if undrained, or CK_0D , if drained.

Generalisation of the critical state model

The critical state model can accommodate *any* stress path, therefore, it is *generalised.* It incorporates the effective stress concept regardless of total stresses and it is *symmetric* in relation to the hydrostatic axis. These key aspects of the model will be explained as follows.

Figure 13.4 presents the undrained behaviour of an *NC* clay. The path in the *s':e* diagram is *A'B'* (figure 13.4c), which is a horizontal line, since no volume change is allowed. Point *A'* belongs to the *ICL*, and point *B'*, to the *CSL*. The corresponding *ESP's* (figure 13.4b) start at point *A* and reach B_1 or B_2 , as the *TSP* is directed upward or downward. The model is symmetric in relation to the abscissae (figure 13.4a).

Figure 13.4b shows that the *ESP's* are unchanged irrespective of the applied *TSP's*. The pore pressure behaviour for each *TSP* will be different, in order to keep effective stresses unchanged. As an example, the u_f value for the axial extension test is given by the distance between points *E* and B_2 . In this case, since $u_f < u_0$, Δu_f is negative. Exactly the opposite occurs in compression tests, in which Δu_f is positive.

In a drained shear test (figure 13.5) the critical state model assumes that the

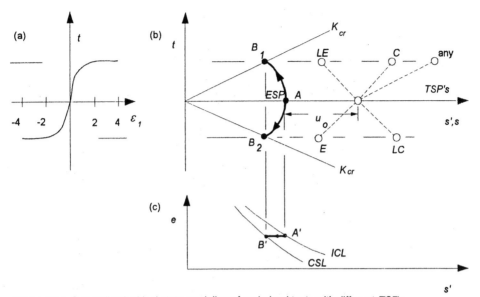

Figure 13.4. Generalised critical state modeling of undrained tests with different *TSP*s

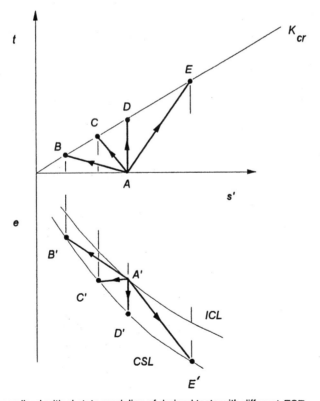

Figure 13.5. Generalised critical state modeling of drained tests with different *ESP*s

Table 13.1. Volume change according to the *ESP* during drained shear (figure 13.5)

ESP	Change in void ratio during shear
AB	increase
AC	decrease
AD	decrease
AE	decrease

sample volume will change until it reaches the critical state. For each *ESP* shown in figure 13.5 the volume change will be different, as indicated in table 13.1.

Interpretation of practical problems in geotechnical engineering

The stress path technique will be used for the interpretation of a few practical problems in geotechnical engineering like retaining walls, embankments and excavations in clay and pile foundation.

Retaining walls

Consider a retaining wall (figure 13.6) with a granular backfill. The wall is assumed frictionless. The stresses acting on the wall can be evaluated through a simple model which forms the basis of the *Rankine earth pressure theory*, described as follows:

– *The active state* (figure 13.6a): Consider a soil element P in the backfill. If the wall moves outward, contrary to the backfill, during the construction, P will suffer a *decrease* in the horizontal stress while the vertical stress does not change. The *ESP* followed by P is AB, corresponding to lateral extension. The final state of stress is called *active state*, corresponding to the horizontal effective stress σ'_{ha}.

– *The passive state* (figure 13.6b): It is the opposite situation. An external loading, such as a bridge foundation, applies a loading that pushes the wall against the backfill. The soil element P suffers an increase in the horizontal stress but the vertical stress remains unchanged. The corresponding *ESP* is AB, lateral compression, in which horizontal stress increase until failure is reached at B. The value of the horizontal stress at this point is σ'_{hp}.

– K_0 *state* (figure 13.6c): The K_0 state occurs during consolidation with no lateral strain. An equivalent situation may be assumed in the vicinity of a very rigid structure which prevents any soil displacements during backfilling. The *ESP* coincides with the K_0 line. Eventually, the horizontal stress will assume an intermediate value σ'_{h0}, between σ'_{ha} and σ'_{hp}.

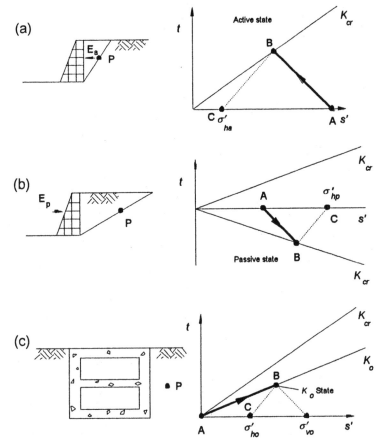

Figure 13.6. Behaviour of a soil element at the backfill of an earth retaining structures. (a) Active state; (b) Passive state; (c) K_0 state

Embankment on soft ground, one stage of construction

Figure 13.7 presents the $s':t:e$ diagram for a point in the soft foundation under the axis of an embankment constructed in one stage. Normal construction time is short enough and undrained condition is assumed. The *ESP* is *AB*, which does not reach the K_{cr} line, because construction stops before failure. The minimum factor of safety occurs at *B*, closest to the K_{cr} line . The *TSP* is *AC*, corresponding to axial compression.

Portions *BC* and *B'C'* correspond to consolidation and pore pressure dissipation after loading. Total stresses are assumed to be constant. During this phase, the *ESP* goes from *B* to *C*, as pore pressures decrease. The factor of safety, on the other hand, increases, as the *ESP* moves away from the K_{cr} line.

Embankment on soft ground constructed in two stages

When loading exceeds the bearing capacity of a soft foundation, the designer's option may be to divide the construction in stages, allowing time for consolida-

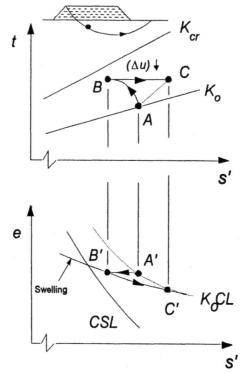

Figure 13.7. Behaviour of a soil element in the foundation of an embankment constructed in one stage on soft soils

tion between them. This leads to a gain in undrained strength in the soft foundation soil.

This case is represented in figure 13.8 for point P under the embankment. The *ESP* at the first loading stage is AB. Loading is halted before failure occurs at C. The initial undrained shear strength is c_{u0}.

Consolidation is allowed, pore pressure dissipates without significant change in total stresses and the *ESP* goes from B to D. Then, construction is resumed and loading is placed again. The *ESP* reaches E in undrained conditions and construction is once more stopped to avoid failure at F. The ordinate of point F gives the final undrained shear strength c_{uf}, higher than the initial value, demonstrating a gain in undrained strength.

Excavation in soft clay

Dredging engineers are aware of the phenomenon of flattening of dredged slopes in soft ground. If a canal in soft soil is dredged with slopes 1:1, it may fail within a few days. The final slopes become, for instance, 1:6 (*V:H*).

This fact can be explained by means of the *s':t:e* diagram presented in figure 13.9. The *TSP* is *AC* corresponding to axial extension in undrained condi-

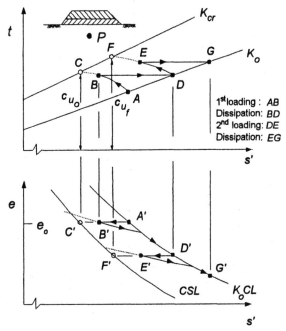

Figure 13.8. Behaviour of of a soil element in the foundation of a staged constructed embankment on soft clay

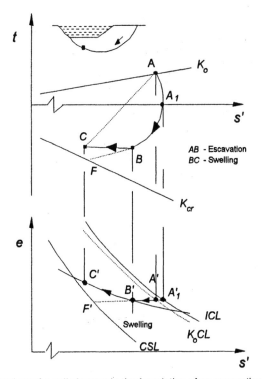

Figure 13.9. Behaviour of a soil element in the foundation of an excavation in soft clay

tions. The *ESP* is AA_1B, halted at *B*, just before failure at *F*, if the excavation went on. Point *C* of the *TSP* is at the left of point *B* of the *ESP*, the pore pressure change was negative. With time, swelling takes place, and pore pressure equilibrium may be reached at *C*. The *ESP* goes from *B* to *C*. As it appraches the K_{cr} line, the safety factor decreases.

Pile in *NC* clay

Pile design methods are very simple in practice and usually employ total stresses. An effective stress approach (e.g., Kraft, 1982) allows an insight into the problem. Our study will concentrate on the behaviour of a soil element *P* (figure 13.10) in contact with the pile side. Undrained behaviour is assumed.

P will be loaded in *lateral compression* during pile installation, the *TSP* is *AB*, reaching the critical state at *B*. After installation time is allowed and dissipation takes place. The total stresses may vary leading to what is known as *pile set up*. This means, a gain in soil strength at *P*, as pore pressure dissipates. At the end of the set up time, the stresses will be at *C*, away from the K_f line.

If the pile is now quickly loaded to failure, the *ESP* is *CD*, reaching once more the critical state at *D*. During loading, the applied total stresses at *P* are essentially shear. Therefore, the values of *q* or *t* will be changed, but keeping the

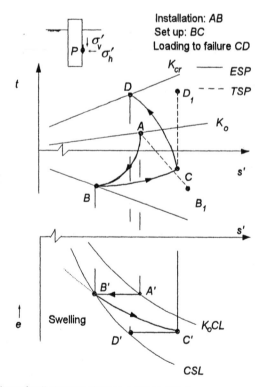

Figure 13.10. Behaviour of soil element close to a pile in *NC* clay

total average stresses p or s essentially constant (Lopes, 1985). The correspondent *TSP* plots vertically, as represented by the segment AD_1.

Pile in *OC* clay

Similar to the previous case, except that the initial stresses and the void ratio at P (figure 13.11), correspond to points A and A', respectively below the K_0 line and on the dry side of the *CSL*.

Residual shear strength

The *residual* shear strength occurs in clayey soils subjected to very large shear deformations, of the order of *metres*. It is caused by landslides on soil elements in the vicinity of the failure surface.

The critical state model is the last condition while soil still behaves as a continuum. The residual state is well beyond that, after a slip occurs. Soil particles in the vicinity of the slip are to be aligned according to the failure plane. This

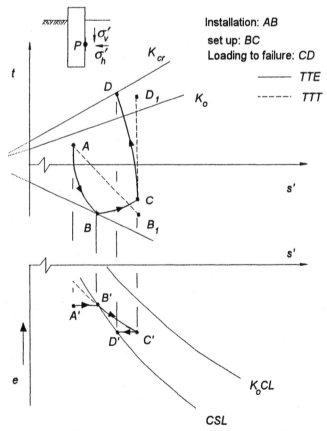

Figure 13.11. Behaviour of soil element close to a pile in *OC* clay

explains why the strength in the residual state is only a fraction of the peak strength.

These concepts are also explained in figure 13.12. Results of an *OC* clay and a dense sand are shown. In sand, the peak strength is due to grain packing, as discussed before in chapter 9. Further straining and dilation, will bring it to the critical state, where there is no longer any effect of packing. If deformation continues beyond that, there is no more change is the shear strength, friction is due to slipping and rolling between grains, which are not affected by the state. It may be concluded that the residual strength in pure sands is equal to the strength at the critical state.

On the other hand, the *OC* clay shows a peak, followed by a critical state condition in which soil is still a continuum and clay particles present a random orientation. Further deformation leads to a slip surface and soil particles will tend to be aligned parallel to the fracture. The shear strength drops until it stabilizes at the residual shear strength.

The residual strength parameters c'_r and ϕ'_r are much lower than the critical state strength parameters, according to the example in table 13.2.

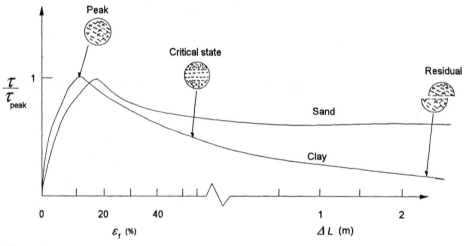

Figure 13.12. Normalised stress-strain curves for dense sand and *OC* clay comparing peak, critical state and residual state

Table 13.2. Example of the difference between strength parameters according to the state of deformation

Effective strength parameters	c' (kPa)	ϕ' (º)
Peak	10	35
Critical state	0	30
Residual	0	17

Residual strength parameters can be evaluated through backanalysis of slope failures, or through laboratory testing in very large deformation conditions, as discussed in detail by Fell and Jeffery (1987). The direct shear with multiple reversions, or the ring shear apparatus, discussed briefly in chapter 8, can be used.

If the shear box is used, the soil specimen is sheared continuously, reaching

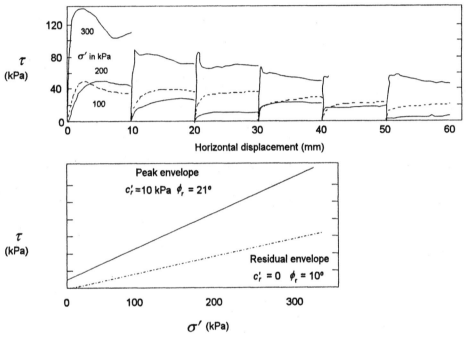

Figure 13.13. Drained direct shear test results with multiple reversions on clay from Curitiba, Brazil (Massad et al., 1981)

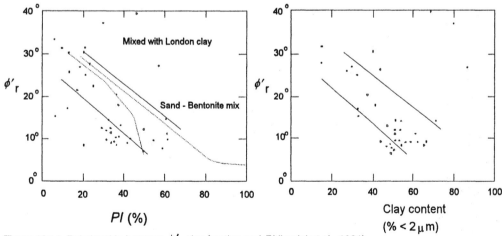

Figure 13.14. Relationship between ϕ'_r, clay fraction and PI (Lupini et al., 1981)

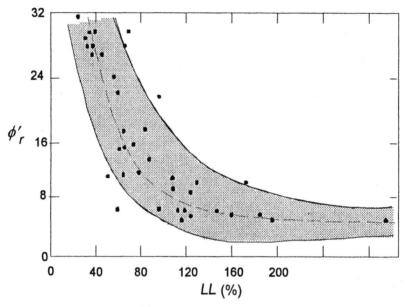

Figure 13.15. Relationship between ϕ'_r and *LL* (Mesri and Cepeda-Diaz, 1986)

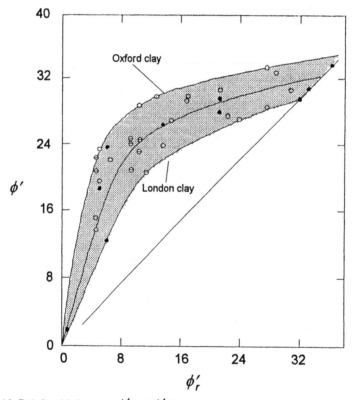

Figure 13.16. Relationship between ϕ'_r and ϕ' (Mesri and Cepeda-Diaz, 1986)

full box displacement in either direction, in a slow zig-zag movement. As an example, figure 13.13 presents data on a residual clay specimen in which the friction angle dropped from 21 to 10° after multiple reversions, and was used for the assessment of slope stability problems (Massad et al., 1981).

Preliminary assessment of ϕ' and ϕ'_r as a function of the clay fraction or Atterberg limits can be done through figures 13.14 to 13.16.

Proposed exercises

13.1. In which cases should the residual strength be used for clays and sands? How to determine it?

13.2. Consider a clay with the following critical state parameters: $\phi' = 33°$, $C_c = 1.3$, $C_s = 0.02$ and $e_{cs} = 6.5$. Work out: (a) the equation of the *CSL*; (b) the void ratio at the critical state for *CID-C*, *CID-E*, *CID-LC*, *CID-LE*, all starting shear at $s' = 150$ kPa; (c) the pore pressure change in the following triaxial tests: *CIU-C*, *CIU-E*, *CIU-LC*, *CIU-LE*, also starting shear at $s' = 150$ kPa.

13.3. Repeat previous exercise for specimens consolidated along the the K_0 line. Assume $K_0 = 0.6$ and the confining stress before shear equals to 200 kPa. Locate the K_0 line in the *s':e* diagram with a given $e_{c0} = 6.1$.

13.4. Consider a soil element in contact with the side of pile installed in *NC* clay. Sketch the *ESP* for this element during the phases of: installation by driving, set-up (pore pressure dissipation) and eventually quick axial loading to failure.

13.5. Repeat previous exercise for *OC* clay.

13.6. Repeat previous exercise for dense sand.

Chapter 14

Cam-clay

Introduction

This chapter presents the theoretical basis for the application of an elasto-plastic model named Cam-clay used for calculation of strains in a soil element for a given change in the state of stress. Cam-clay incorporates the critical state theory plus a series of plasticity concepts, which will be briefly discussed. Equations will be presented, but not deduced.

A computer program named *Cris* is distributed with the book for hands-on training on critical state models. The program performs simulation of triaxial tests results.

The reader interested in a deeper and a comprehensive review of plasticity and the deduction of Cam-clay equations should refer to the following books: Desai and Siriwardane (1984), Britto and Gunn (1987), Bolton (1979) and Schofield and Wroth (1968).

The elasto-plastic model

Strains in an elasto-plastic model can be treated separately in the elastic and in the plastic domain. Figure 14.1 presents a *e:log s'* diagram in which a soil element deforms from point A to C. Path AC is divided into AB and BC. The first, lies on the swelling line, the second, is vertical. Strains along the swelling line were discussed in chapter 6. They are small and reversible and, therefore, *elastic*. On the other hand, the strains along BC are irreversible and, thus, *plastic*.

The energy spent, or the work W, during the deformation of a soil element is given by (Timoshenko and Goodier, 1951):

$$W = \sigma_1' \varepsilon_1 + \sigma_2' \varepsilon_2 + \sigma_3' \varepsilon_3 \tag{14.1}$$

In the elastic domain this energy is stored by the soil, while in the plastic, part of this energy is dissipated as heat. Hence:

$$W = W_{stored} + W_{dissipated} \tag{14.2}$$

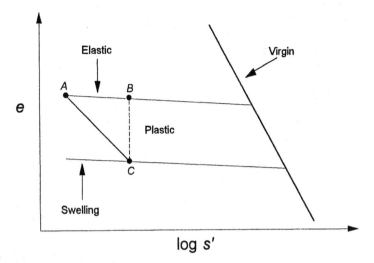

Figure 14.1. Decomposing the volumetric strains in elastic and plastic strains

The elasto-plastic models differ according to the assumption made for the energy dissipation during the plastic regime.

Yielding

Stress-strain models for a soil element are shown in figure 14.2. Soil deforms elastically until it reaches point E, where *yielding* starts. This means that, in addition to elastic strains ε^e, the material also deforms plastically by an amount ε^p, the total strain ε being the sum:

$$\varepsilon = \varepsilon^e + \varepsilon^p \tag{14.3}$$

where: superscripts ^{e} and ^{p} indicate, respectively, elastic and plastic strains.

As yielding begins, the soil element can behave in three ways. In the first (figure 14.2a), the stress remains constant and the element continues to deform. It is an *elastic-perfectly-plastic* behaviour. In the second, (figure 14.2b), in which the strength is reduced as strains increase, is known as *strain softening*. Finally (figure 14.2c), in the third one, strength increases with deformation; it is called *strain hardening*.

A soil element starts to yield when its effective stress path touches a particular convex surface defined in the stress space $p':q$ or $s':t$ known as the *yield locus* (figure 14.3). Below this surface, all strains are assumed to be purely elastic, as shown in figure 14.4. As the effective stress path AE touches the yield locus, plastic strains will be added to the elastic strains.

Elasto-plastic models differ according to the shape assumed for the yield locus. Two slightly different types will be covered in this chapter. One, is the *Cam-clay*, another, the *Modified Cam-clay*. Both can be used through the *Cris* program.

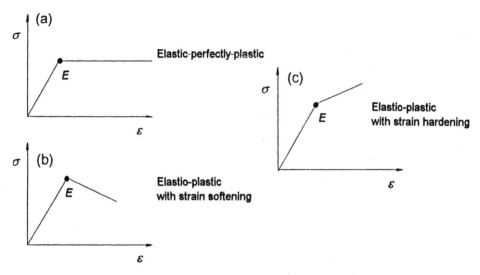

Figure 14.2. Different patterns of stress-strain curves considered in elasto-plastic models

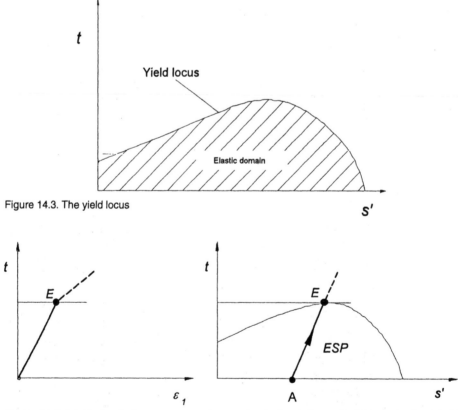

Figure 14.3. The yield locus

Figure 14.4. (a) Stress-strain curve presenting purely elastic strains until yield starts at point *E*; (b) corresponding yielding locus

The *s':t:e* and the *p':q:e* diagrams

The equations of the Cam-clay models are now presented for the *p':q:e* diagram, instead of than the *s':t:e*, used in this book. Slightly different notations are utilized. However, it is easy to move from one set to the other.

The *s':t:e* and *p':q:e* diagrams in figures 14.5 and 14.6, tables 14.1 and 14.2 compare the assumptions and equations utilized by the models.

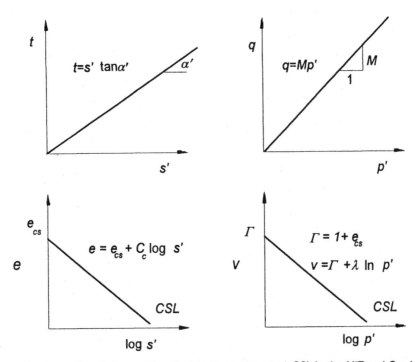

Figure 14.5. Comparison between the critical state envelope and *CSL* in the MIT and Cambridge plots

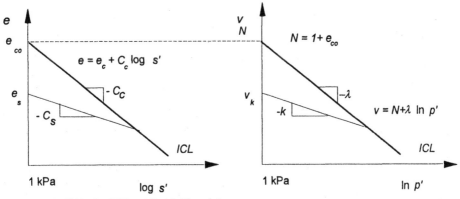

Figure 14.6. *ICL* in the MIT and Cambridge plots

Table 14.1. Parameters used in the *s':t:e* and *p':q:e* (or *v*) diagrams

Name	*s':t:e* diagram	*p':q:e* diagram	Relationship between parameters	
	s'	p'	$p' = \dfrac{1}{3}(3s' - t)$	(14.34)*
Variables	t	q	$q = 2t$	(14.35)
	e	e *(or v)*	$v = 1 + e$	(14.36)**
Strength envelope	$\tan \alpha'$	M	$M = \dfrac{6\sin\phi'}{3 - \sin\phi'}$	(14.37)
CSL	e_{cs}	Γ	$\Gamma = e_{cs} + 1$	(14.38)
	C_c	λ	$\lambda = C_c/2.3$	(14.39)
ICL	e_c	N	$N = e_c + 1$	(14.40)
Swelling line	e_s	V_k	$V_k = e_s + 1$	(14.41)
	C_s	κ	$\kappa = C_s/2.3$	(14.42)

*   Axi-symmetric condition: $\sigma'_2 = \sigma'_3$
** *v* is the specific volume

Table 14.2. Equations in the diagrams *s':t:e* and *p':q:e* (or *v*)

Name	*s':t:e* diagram	*p':q:e* (or *v*) diagram	
Strength envelope	$t = s' \tan\alpha'$	$q = M\,p'$	(14.13)
CSL	$e = e_{cs} + C_c\,\log s'$	$V = \Gamma + \lambda \ln p'$	(14.14)
ICL	$e = e_c + C_c\,\log s'$	$V = N + \lambda \ln p'$	(14.15)
Swelling line	$e = e_s + C_s\,\log s'$	$V = V_k + \lambda \ln p'$	(14.16)
ESP line	$\kappa = \sigma'_3/\sigma'_1$	$\eta = q/p'$	(14.17)

Yield locus

The equation for the yield locus in the Cam-clay model is:

$$q = M\,p' \ln\left(p'_m/p'\right) \tag{14.18}$$

It is presented graphically in figure 14.7a. Parameter p'_m corresponds to the iso-tropic pre-consolidation stress and also the interception between the yield locus and the hydrostatic axis.

Cam-clay can be considered a big step in the theoretical formulation of soil behaviour. It is a powerful and simple tool. The original model was improved by Burland (1967) and is called *Modified Cam-clay*. He employed an elliptical yield

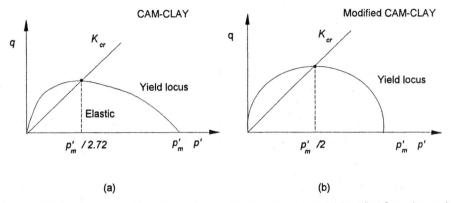

Figure 14.7. Different assumptions for yield locus: (a) Cam-clay model; (b) Modified Cam-clay model

locus (figure 14.7b) given by:

$$M^2 p'^2 - M^2 p'_m p' + q^2 = 0 \tag{14.19}$$

rearranging the terms:

$$q = M p' \sqrt{\frac{p'_m}{p'} - 1} \tag{14.20}$$

Strains

Starting from equation 14.1, it is possible (e.g., Schofield and Wroth, 1968) to relate strain energy to values of stress invariants p' and q, through:

$$W = q \varepsilon_s + p' \varepsilon_v \tag{14.21}$$

where: ε_s and ε_v are defined as shear and volumetric strains, given by:

$$\varepsilon_s = \frac{2}{3} \left(\varepsilon_1 - \varepsilon_3 \right) \tag{14.22}$$

$$\varepsilon_v = \left(\varepsilon_1 + 2\varepsilon_3 \right) \tag{14.23}$$

where: ε_1 and ε_3 are principal strains.

The values of ε_s and ε_v can be decomposed as a sum of elastic and plastic strain components. Therefore:

$$\varepsilon_s = \varepsilon_s^e + \varepsilon_s^p \tag{14.24}$$

$$\varepsilon_v = \varepsilon_v^e + \varepsilon_v^p \tag{14.25}$$

Similar equations are obtained for elementary strain increments by differentiation:

$$d\varepsilon_s = d\varepsilon_s^e + d\varepsilon_s^p \tag{14.26}$$

$$d\varepsilon_v = d\varepsilon_v^e + d\varepsilon_v^p \tag{14.27}$$

In the Cam-clay and Modified Cam-clay models strains are obtained as follows:

Elastic phase

$d\varepsilon_v^e$ is the elastic volumetric strain increment, obtained by differentiation of the equation of the swelling line (equation 14.16), for an infinitesimal change in the p' value. It is given by:

$$d\varepsilon_v^e = \frac{\kappa}{1+e} \frac{dp'}{p'} \tag{14.28}$$

$d\varepsilon_s^e$ is the *elastic shear strain increment*, obtained from Hooke's law:

$$d\varepsilon_s^e = \frac{dq}{3G} \tag{14.29}$$

where: G is the shear modulus.

Plastic phase

Plastic strains are obtained from what is called in plasticity the strain hardening law, which in the Cam-clay model is:

$$d\varepsilon_v^p = \frac{de^p}{1+e} = \frac{\lambda - \kappa}{1+e} \left(\frac{dp'}{p'} + \frac{d\eta}{M} \right) \tag{14.30}$$

The value of the plastic shear strain increment de is obtained from what is called in plasticity the *flow rule*, which for the Cam-clay model is:

$$\frac{d\varepsilon_s^p}{d\varepsilon_v^p} = \frac{1}{M - \eta} \tag{14.31}$$

The equivalent equations in the Modified Cam-clay model are:

$$d\varepsilon_v^p = \frac{de^p}{1+e} = \frac{\lambda - \kappa}{1+e} \left(\frac{dp'}{p'} + \frac{2\eta \, d\eta}{M^2 + \eta^2} \right) \tag{14.32}$$

$$\frac{d\varepsilon_s^p}{d\varepsilon_v^p} = \frac{2\eta}{M^2 + \eta^2} \tag{14.33}$$

Automatic simulation of triaxial tests

As a part of the study programme on critical state, we now introduce the *Cris* training tool. *Cris* is a computer program to predict triaxial test results and to plot the diagrams $p':q:e$ or $s':t:e$. It is based on the previous work by Almeida et al. (1987) with several improvements.

Cris was written in QuickBasic, being distributed in executable format for use in IBM-PC microcomputers. It was designed to be self-explanatory and user-friendly. Some tips on its use are presented below.

Input data

Critical state requires only five parameters ϕ', C_c, C_s, G and e_{cs} and *Cris* asks the user to assign values for each of them. The user inputs soil data trough the keyboard or trough a previous file created by the program. For the inexperienced user, *Cris* has default data for a soft clay. Just press *Enter* on the keyboard. On a second run you can alter the default values.

Critical state parameters ϕ', C_c, C_s, G and e_{cs} are obtained from triaxial tests and / or estimated through the empirical correlations presented in this and in the previous chapters.

As an example, input data will be estimated for a high plastic clay. ϕ' can be evaluated from figure 10.8 as a function of the plasticity index *PI*. The empirical correlations presented in chapter 6 allow the evaluation of the compression in-

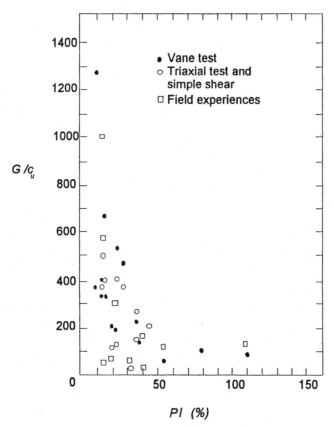

Figure 14.8. Relationship between G / c_u and *PI* (Holtz and Kovacs, 1981)

Table 14.3. Summary of critical state parameters for soft clay used as default values in the *Cris* program

Critical state parameter	Value
ϕ'	30º
C_c	2
C_s	0.3
G	2000 kPa
e_{cs}	5

dex C_c. The value of the swelling index C_s can be taken as a percentage of C_c, of the order of 10 to 20%.

The shear modulus G for clays can be evaluated through the ratio G/c_u, which was found to vary with the plasticity index *PI*, as shown in figure 14.8 (Holtz and Kovacs, 1981). For a *PI* value between 50 and 100%, it follows that $G/c_u = 200$. Taking, for instance, $c_u = 10$ kPa, then: $G = 2000$ kPa.

The value of e_{cs} can be obtained from *CIU* triaxial tests extrapolating the *CSL* for $p' = 1$ kPa. Alternatively, oedometer tests can be employed.

The critical state parameters employed in this example are summarized in table 14.3.

Test definition

The program will ask the test conditions and how the user wishes to plot the results. The questions are:

– The output diagram, either the Cambridge $p':q$ plot or the or the MIT's $s':t$. The latter was set as default.

– The initial pre-shear value of the mean stress (p' or s'): a default value of 150 kPa was assigned.

– The value of s' (or p') corresponding to the point of interception of the yield locus with the abscissae. This value can be figured out from the overconsolidation mean stress of the soil. A default value of 200 kPa was assigned.

– Drainage conditions: Either drained or undrained tests can be simulated. The latter is the default choice.

– The slope (ds/dt) of the *TSP*. A default value of $ds/dt = 1$ (45°) was assigned. This corresponds, in the Cambridge plot, to $dq/dp = 3$.

– The value of the shear strain increment $d\varepsilon_s$ to be used in each step in the calculations. The selected value should be small in order to avoid numerical problems. The default value is 0.2%.

– The test type, either compression or extension one, the first being selected as default.

– The Cam-clay model: The user can choose between the original Cam-clay or the Modified Cam-clay. The latter was chosen as default.

Results

The results are presented in the following ways:

Output data: All output numerical data are stored in a ASCII file named *Cris.out* (table 14.4) located in the current directory. All alphabetic characters in this file are written between quotation marks ("). This enables the file to be directly imported by spreadsheet programs, like *Lotus* or *Quattro*, for additional processing and high resolution plotting.

Screen graphics: The main plots are presented on the screen. The program asks which type of plot is desired. If the option *All* is selected, the results are shown as indicated in figure 14.9. On the top, the *s':t* diagram and the stress-

Table 14.4. Cris output data, *CIU* test

```
        "Program CRIS"
"Units :  stresses, pore pressure, shear modulus- kPa"
     "strains - %"

"Soil Properties"
   "Fi = 30        Mu = 1.2 "
   "Cc  = 2        Lambda = .8695652 "
   "Cs = .3        Kappa = .1304348 "
   "G = 2000 "
   "ecs = 5 "

"Test type :  undrained, compression"

"Initial stresses :"
   "s' = 150     t = 0 "

"Intersection of the yield curve with the (s' or p') axis"
   "s'm = 200 "

"Slope of the total stress path :  dt/ds = .4285714 "

"Strain increment (dEs):  .2 "

"Model :  Cam-clay modified"

"Final conditions :"
   "s' = 127.52    t = 63.76    du = 86.24    A =  0.69"
   "v = 1.943      e = 0.943"
```

"Es(%)"	"E1"	"ds"	"s'"	"dt"	"t"	"du"	
0.0000	0.0000	17.3205	150.0000	51.9615	0.0000	0.0000	"Elas"
0.8660	0.8660	17.3205	167.3205	51.9615	51.9615	34.6410	"Plas"
1.0660	1.0660	-1.1900	165.2172	0.4113	53.1298	37.9127	"Plas"
1.2660	1.2660	-1.0886	163.2290	0.3537	54.1539	40.9249	"Plas"
1.4660	1.4660	-0.9992	161.3535	0.3064	55.0550	43.7015	"Plas"
1.6660	1.6660	-0.9198	159.5864	0.2672	55.8507	46.2643	"Plas"
1.8660	1.8660	-0.8491	157.9225	0.2344	56.5559	48.6334	"Plas"
2.0660	2.0660	-0.7858	156.3563	0.2068	57.1830	50.8268	"Plas"
2.2660	2.2660	-0.7289	154.8821	0.1833	57.7426	52.8605	"Plas"
2.4660	2.4660	-0.6775	153.4942	0.1632	58.2435	54.7493	"Plas"
2.6660	2.6660	-0.6309	152.1872	0.1459	58.6931	56.5060	"Plas"
2.8660	2.8660	-0.5886	150.9557	0.1310	59.0979	58.1422	"Plas"
3.0660	3.0660	-0.5501	149.7950	0.1180	59.4634	59.6683	"Plas"

Table 14.4. Continued

"Es(%)"	"E1"	"ds"	"s'"	"dt"	"t"	"du"	
3.2660	3.2660	-0.5148	148.7004	0.1066	59.7941	61.0937	"Plas"
3.4660	3.4660	-0.4824	147.6674	0.0966	60.0941	62.4267	"Plas"
3.6660	3.6660	-0.4527	146.6922	0.0877	60.3670	63.6747	"Plas"
3.8660	3.8660	-0.4253	145.7710	0.0799	60.6156	64.8447	"Plas"
4.0660	4.0660	-0.4000	144.9002	0.0730	60.8427	65.9425	"Plas"
4.2660	4.2660	-0.3766	144.0767	0.0668	61.0504	66.9737	"Plas"
4.4660	4.4660	-0.3549	143.2975	0.0613	61.2409	67.9434	"Plas"
4.6660	4.6660	-0.3348	142.5597	0.0563	61.4158	68.8560	"Plas"
4.8660	4.8660	-0.3160	141.8610	0.0518	61.5766	69.7157	"Plas"
5.0660	5.0660	-0.2986	141.1988	0.0478	61.7249	70.5261	"Plas"
5.2660	5.2660	-0.2823	140.5709	0.0442	61.8617	71.2908	"Plas"
5.4660	5.4660	-0.2671	139.9754	0.0409	61.9881	72.0127	"Plas"
5.6660	5.6660	-0.2529	139.4102	0.0379	62.1051	72.6949	"Plas"
5.8660	5.8660	-0.2395	138.8737	0.0351	62.2135	73.3398	"Plas"
6.0660	6.0660	-0.2270	138.3642	0.0326	62.3142	73.9500	"Plas"
6.2660	6.2660	-0.2153	137.8801	0.0304	62.4076	74.5276	"Plas"
6.4660	6.4660	-0.2043	137.4199	0.0283	62.4946	75.0747	"Plas"
6.6660	6.6660	-0.1939	136.9824	0.0264	62.5756	75.5932	"Plas"
6.8660	6.8660	-0.1842	136.5663	0.0246	62.6511	76.0847	"Plas"
7.0660	7.0660	-0.1750	136.1704	0.0230	62.7215	76.5511	"Plas"
7.2660	7.2660	-0.1663	135.7937	0.0216	62.7874	76.9937	"Plas"
7.4660	7.4660	-0.1581	135.4350	0.0202	62.8489	77.4139	"Plas"
7.6660	7.6660	-0.1504	135.0935	0.0189	62.9065	77.8130	"Plas"
7.8660	7.8660	-0.1431	134.7682	0.0178	62.9605	78.1923	"Plas"
8.0660	8.0660	-0.1362	134.4583	0.0167	63.0111	78.5528	"Plas"
8.2660	8.2660	-0.1297	134.1629	0.0157	63.0586	78.8956	"Plas"

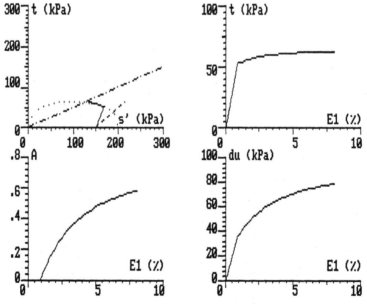

Figure 14.9. *Cris* output graphs for an undrained triaxial test: (a) top left, *s':t* diagram; (b) top right, stress-strain curve; (c) bottom, left, pore pressure parameter *A* versus major principal strain; (d) bottom right, Δ*u* versus major principal strain

strain curve t: ε_1. Below, Skempton's pore pressure parameter A versus strain. These graphs can, then, be plotted through the DOS function *Print Screen*. After printing, press the space bar to return to the menu.

The results of a simulation of a drained *CID* test with the same parameters and under the same confining stress is presented in table 14.5 and in figure 14.10.

Table 14.5. Cris output data, *CID* test

"Program CRIS"

"Units : stresses, pore pressure, shear modulus- kPa"
"strains - %"

"Soil Properties"
"Fi = 30   Mu = 1.2 "
"Cc = 2   Lambda = .8695652 "
"Cs = .3   Kappa = .1304348 "
"G = 2000 "
"ecs = 5 "

"Test type : drained, compression"

"Initial stresses :"
"s' = 150   t = 0 "

"Intersection of the yield curve with the (s' or p') axis"
"s'm = 200 "

"Slope of the total stress path : dt/ds = .4285714 "

"No of increments 20 "

"Model : Cam-clay modified"

"Final conditions :"
"s' = 300.00   t = 150.00   du = 0.00   A = 0.00"
"v = 1.199   e = 0.199"

"s"	"t"	"dEse"	"dEsp"	"dEs"	"Es"	"dEvs"	"dEvp"	"E1"	"dEv"	"Ev"	"e"	
150.000	0.000	0.003	0.000	0.260	0.000	0.002	0.000	0.000	0.233	0.000	0.943	"Elas"
157.799	7.799	0.003	0.000	0.260	0.260	0.002	0.000	0.338	0.225	0.233	0.938	"Elas"
165.599	15.599	0.003	0.000	0.260	0.520	0.002	0.000	0.673	0.219	0.458	0.934	"Elas"
173.398	23.398	0.003	0.000	0.260	0.780	0.002	0.000	1.006	0.212	0.677	0.929	"Elas"
181.198	31.198	0.003	0.000	0.260	1.040	0.002	0.000	1.336	0.206	0.889	0.925	"Elas"
"Begin of elasto - plastic behaviour ..."												
188.997	38.997	0.003	0.019	2.192	1.300	0.002	0.027	1.665	2.911	1.095	0.921	"Plas"
196.796	46.796	0.003	0.025	2.766	3.491	0.002	0.028	4.827	3.047	4.006	0.866	"Plas"
204.596	54.596	0.003	0.031	3.409	6.257	0.002	0.029	8.608	3.149	7.053	0.809	"Plas"
212.395	62.395	0.003	0.039	4.138	9.667	0.002	0.030	13.067	3.226	10.202	0.752	"Plas"
220.195	70.195	0.003	0.047	4.980	13.805	0.002	0.031	18.281	3.284	13.429	0.695	"Plas"
227.994	77.994	0.003	0.057	5.974	18.785	0.002	0.031	24.356	3.326	16.712	0.640	"Plas"
235.793	85.793	0.003	0.069	7.179	24.759	0.002	0.032	31.439	3.357	20.038	0.585	"Plas"
243.593	93.593	0.003	0.084	8.687	31.938	0.002	0.032	39.737	3.380	23.395	0.532	"Plas"
251.392	101.392	0.003	0.104	10.651	40.625	0.002	0.032	49.551	3.397	26.775	0.480	"Plas"
259.192	109.192	0.003	0.131	13.336	51.276	0.002	0.032	61.334	3.409	30.172	0.430	"Plas"
266.991	116.991	0.003	0.170	17.262	64.612	0.002	0.032	75.806	3.419	33.581	0.381	"Plas"
274.790	124.790	0.003	0.233	23.588	81.875	0.002	0.032	94.208	3.428	37.001	0.334	"Plas"
282.590	132.590	0.003	0.353	35.552	105.463	0.002	0.032	118.939	3.436	40.428	0.288	"Plas"
290.389	140.389	0.003	0.666	66.894	141.015	0.002	0.032	155.636	3.443	43.864	0.244	"Plas"

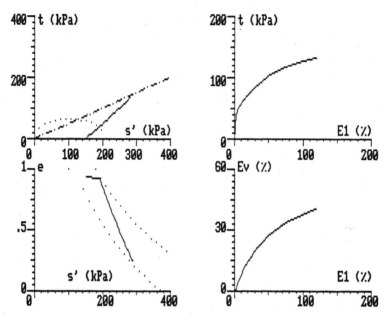

Figure 14.10. *Cris* output for a drained triaxial test: (a) top left, *s':t* diagram; (b) top right, stress-strain curve; (c) bottom, left, *s':e* diagram, showing the *ICL* and the *CSL*; (d) bottom right, volumetric strains versus major principal strain

Proposed exercises

14.1. Obtain critical state parameters for the Rio de Janeiro clay from the *CIU* test data in figure 11.2 and the *e:log p'* data in figure 11.18. The steps are: (a) extrapolate by sketching in figure 11.2 the pore pressure and strength curves until critical state reached; (b) work out data points from the figure and use a spreadsheet program to plot the results; (c) still in the spreadsheet, plot the MIT and the Cambridge diagrams and work out values of ϕ' and G; (d) from figure 11.18, obtain C_c and e_{cs} for the stress range of the *CIU* test in figure 11.3 ($\cong$ 150 kPa) and evaluate C_s.

14.2. Enter the parameters obtained in the previous exercise into the *Cris* program to simulate the triaxial test data in figure 11.3. Use the Modified Cam-clay model. Compare the results with experimental data and then slightly modify each parameter, one at a time, and observe the effect. Can the match be improved?

14.3. Repeat previous exercise for the original Cam-clay model. Can you observe any improvement in the match?

14.4. Use the critical state parameters obtained in exercise 14.1 to simulate a drained triaxial test on *NC* clay under a confining stress of 150 kPa.

SI units in geotechnics

Quantity	Symbol	Name	Usual multiples and sub-multiples
Length	m	metre	km, cm, mm, μm[*]
Mass	kg	kilogram	g, Mg
Force, weight	N	Newton	kN, MN
Pressure or stress	Pa	Pascal[**]	kPa, MPa, GPa
Unit mass	kg/m^3	kilogram per cubic metre	
Unit weight	kN/m^3	kilonewton per cubic metre	
Density	pure number		
Time	s	second	min, h
Coefficient of consolidation	$m^2/year$	square metre per year	
Coefficient of permeability	m/s	metre per second	

[*] μm is the *micrometre* and it is equal to 10^{-6} of a metre
[**] 1 Pa=1 N/m^2

Unit	Symbol	
	Right	Wrong
Length	m, km, μm	Km, μ
Mass	g, kg, Mg	gr, Kg, ton
Force, weight	N, kN, MN	kgf, tf, KN, kn
Pressure, stress	kPa, MPa, GPa	kgf/cm2, tf/m^2, KPa
Unit mass	kg/m^3, Mg/m^3	t/m^3
Unit weight	kN/m^3	kgf/m^3, tf/m^3
Time	h, min, s	sec, seg

Laboratory tests standards

Tests	ABNT	ASTM	BS	CESP	DNER	USBR
Water content		D 2216	1377	MSL-01	E-9	
LL	NBR 6459	D 423	1377	MSL-04	ME 44-71	E-7
PL	NBR 7180	D 424	1377	MSL-04	ME 82-63	E-7
G_s	NBR 6508	D 854	1377	MSL-03	ME 93-64	E-10
Grain size analysis	NBR 7181	D 422	1377	MSL-05	E18	E-6
Relative density	MB 3324	D 2049	1377	MSL-11	ME 93-64	E-12
Drained direct shear			USCS (1970)			
Consolidation	MB 3326	D 2435	1377		IE 05-71	E-17
UU triaxial Head (1986)		D 2850	1377			E-17
CIU triaxial Head (1986)						E-17
CID triaxial Head (1986)						
Permeability constant head		D 2434		MSL-09		
Permeability variable head				MSL-09		

ABNT - Associação Brasileira de Normas Técnicas, ABGE - Associação Brasileira de Geologia de Engenharia, ASTM - American Society for Testing and Materials, BS - British Standards Institution, CESP - Companhia Energética de São Paulo, DNER - Departamento Nacional de Estradas de Rodagem, USBR - US Bureau of Reclamation.

In situ tests standards

Tests	ABNT	ASTM	BS	USBR
SPT	NBR 6484	D4633		E21
CPT & CPTU	MB 3406			
Field vane test	NBR 10905	D2573	BS1377	E20

International Standards for CPT, SPT and other in situ tests were published by the Technical Committee on Penetration Testing of Soils - TC 16 of the ISSMFE International Society for Soil Mechanics and Foundation Engineering. The final report was published by the Swedish Geotechnical Institute, Linköping, June, 1989.

Answers to selected exercises

Chapter 1

1.1. $w = 32\%$; $e = 0.85$; $\gamma_d = 14.3$ kN/m³; $\gamma_{sat} = 18.9$ kN/m³; $\gamma_{sub}=8.9$ kN/m³.

1.2. $w = 32\%$; $e = 1.6$; $\gamma_d = 19.2$ kN/m³; $\gamma_{sat} = 25.4$ kN/m³; $\gamma_{sub}=15.4$ kN/m³.

1.3. $\gamma_d = 18$ kN/m³; $\gamma_{sub}=11.3$ kN/m³.

1.6. $e = 0.66$.

1.7. $\gamma_d = 14.6$ kN/m³; $e = 0.84$.

Chapter 2

2.8. $\sigma_\theta = 137$ kPa; $\tau_\theta = 145$ kPa; $\sigma_1 = 260$ kPa in a plane 10° inclined with horizontal; $\sigma_3 = 37$ kPa in a plane 10° inclined with vertical; $\tau_{max} = 147$ kPa in a plane 55° inclined with vertical.

2.9. $\sigma_\theta = 137$ kPa; $\tau_\theta = 145$ kPa; $\sigma_1 = 260$ kPa in a plane 20° inclined with horizontal; $\sigma_3 = 37$ kPa in a plane 10° inclined with 20° vertical; $\tau_{max} = 147$ kPa in a plane 25° inclined with vertical.

2.10. $\gamma_{max} = 15\%$.

Chapter 4

4.6. Point (0,0) = 560 kPa; point (0,15) = 490 kPa; point (6,0) = 528 kPa; point (10,25) = 416 kPa.

Chapter 5

5.4. $u_a = 266$ kPa; $u_b = 236$ kPa; $u_c = 266$ kPa; $u_d = 206$ kPa; $Q_l = 1.6 \times 10^{-5}$ m³/s/m; $i = 0.24$.

5.6. $Q_l = 1.125 \times 10^{-6}$ m³/s/m; $u_p = 85.2$ kPa.

5.7. $Q_l = 5.33 \times 10^{-5}$ m³/s/m.

5.8. $WL_1, \sigma_v = 66$ kPa, $u_0 = 20$ kPa; $\sigma'_v = 46$ kPa, $WL_2, \sigma_v = 89$ kPa, $u_0 = 55$ kPa, $\sigma'_v = 34$ kPa, $v = 1.75 \times 10^{-7}$ m/s.

5.9. $F_p = 5$ kN.

5.10. $F_p = 0.1$ kN.

Chapter 6

6.5. $\sigma'_{vm} = 251$ kPa; $C_c = 0.21$; $C_s = 0.05$; $CR = 11\%$, $SR = 2.5\%$, $p = 0.41$ m.

6.6. $\rho_{total} = 1.9$ m.

Chapter 7

7.5. Log method: $c_v = 1.5 \mathrm{m}^2/\mathrm{year}$, $\sqrt{t}$ method: $c_v = 2.3 \mathrm{~m}^2/\mathrm{year}$; with $c_v = 1.4$ $\mathrm{m}^2/\mathrm{year}$, $k = 1.4 \times 10^{-9}$ m/s, with $c_v = 2.3 \mathrm{~m}^2/\mathrm{year}$, $k = 2.2 \times 10^{-9}$ m/s.

7.6. $\rho_{total} = 2.6$ m.

7.8. $k = 4.10 \times 10^{-10}$ m/s.

Chapter 9

9.3. $\phi' = 19.3°$

9.4. $\phi' = 38.2°$

9.5. CP1: $E_0 = 27$ MPa. $E_{50\%} = 15$ MPa; CP2: $E_0 = 285$ MPa, $E_{50\%} = 172$ MPa; CP1: $v_0 = 0.5$, $v_{50\%} = 0.57$; CP2: $v_0 = 0.85$, $v_{50\%} = 0.48$; CP1: $\phi_{rut} = 44.9°$, $\phi_{crit} = 37.9°$, CP2: $\phi_{rut} = 36.9°$, $\phi_{crit} = 36.5°$; CP1 is a dense sand and CP2 is a loose sand.

9.6. (a) ϕ' between 30° and 32°; (b) ϕ' between 27° and 30°; (c) ϕ' between 33° and 35°.

Chapter 10

10.4. Kenney, $\phi' = 21.3°$; Mayne, $\phi' = 22.5°$.

Chapter 11

11.4. $\phi'_{crit} = 33.7°$, $\phi'_{rut} = 22.6°$; $A_{crit} = 10.4$, $A_{rut} = 0.73$; $\alpha_{crit} = 0.28$, $\alpha_{rut} = 0.50$; $E_u = 10,750$ kPa, $v_u = 0.50$.

11.5. $\alpha = 0$.

11.6. $\phi' = 33.7°$; $\phi'_{rut} = 19.5°$; $\phi'_{crit} = 34.8°$, $A_{crit} = 1$; $\alpha_{crit} = 0.5$.

Chapter 12

12.4. $c_u = 7.5$ kPa.

12.6. (a) *UU*; (b) *CD*; (c) ring shear *CD* test or a multiple reversions *CD* direct shear test; (d) *UU*; (e) *CD*; (f) *UU*.

Chapter 13

13.2. (a) $e = 5.5 - log~s'$, (b) axial compression $e = 2.23$, axial extension $e = 2.23$, lateral compression $e = 2.93$, lateral extension $e = 2.93$; (c) axial compression $\Delta u = 127$ kPa, axial extension $\Delta u = 127$ kPa, lateral compression $\Delta u = 70$ kPa, lateral extension $\Delta u = 70$ kPa; (d) axial compression $e = 2.36$, axial extension $e = 3.06$, lateral compression $e = 2.09$, lateral extension $e = 2.78$; (e) axial compression $s' = 130$ kPa, extension axial $s' = 45$ kPa, lateral compression $s' = 280$ kPa, lateral extension $s' = 95$ kPa.

References

Aas G., Lacasse S., Lunne T. & Höeg K. (1986). Use of in situ tests for foundation design. *Proc. ASCE Conf. on In Situ Tests in Geotechnical Engineering, In Situ '86, Virginia Tech, Blacksburg.* Geotechnical Publication no. 6, pp 1-30.

Al Hussaini M.M. & Townsend F.C. (1975). Investigation of K_0 testing in cohesionless soils. Technical Report S-75-11, US Army Engineers Waterways Experimental Station, Vicksburg, MI, 70 pp.

Almeida M.S.S. (1982). The undrained behaviour of the Rio de Janeiro clay in the light of critical state theories. *Solos e Rochas,* vol. 5:2, pp 3-24.

Almeida M.S.S., Danziger F.A.B. & Oliveira E.B. (1987). Comportamento tensão-deformacão de ensaios triaxiais de solo previsto através de modelos de estados críticos. *Anais 8° Congresso Latino-Americano. e Ibérico sobre Métodos Computacionais para Engenharia, PUC-RJ,* pp 215-233.

Almeida M.S.S., Collet H.B., Ortigao J.A.R. & Terra B.R. (1989). Settlement analysis of embankment on Rio de Janeiro clay with vertical drains. Supplementary contributions by the Brazilian Society for Soil Mechanics to the 12th ICSMFE, Rio de Janeiro, pp 105-110.

Almeida M.S.S., Oliveira W.L., Medeiros C.J. & Porto E.C. (1987). Laboratory tests and design parameters for offshore piles in calcareous soils. *Proc. Int. Symp. on Offshore Engineering, Brazil Offshore,Rio de Janeiro.*

Andrade R.M. (1983). *O controle da subpressão pela drenagem horizontal.* Engevix SA, Rio de Janeiro, 251 pp.

Asaoka A. (1978). Observational procedure of settlement prediction. *Soils and Foundations,* vol. 18:4, pp 87-101.

Atkinson J.H. & Bransby P.L. (1978). *The mechanics of soils – An introduction to critical state soil mechanics.* McGraw-Hill, London, 375 pp.

Azzouz A.M., Baligh M.M. & Ladd C.C. (1983). Corrected field vane strength for embankment design. *ASCE Journal of Geotechnical Engineering,* vol. 109:5, pp 730-733.

Azzouz A.S., Krizek R.J. & Corottis R.B. (1976). Regression analysis of soil compressibility. *Soils and Foundations,* vol. 16:2, pp 19-29.

Barton N. & Kjaernskli B. (1981). Shear strength of rockfill. *ASCE Journal of the Geotechnical Engineering Division,* vol. 107, GT7, pp 873-891.

Bishop A.W. & Bjerrum L. (1960). The relevance of the triaxial test to the solution of stability problems. *ASCE Conf. on Shear Strength of Cohesive Soils, Boulder, Colorado,* pp 437-501.

Bishop A.W. & Henkel D.J. (1962). *The measurement of soil properties in the triaxial test.* Edward Arnold, London, 227 pp.

289

Bjerrum L. (1973). Problems of soil mechanics and construction on soft clays and structurally unstable soils. *Proc. 8th ICSMFE Int. Conf. on Soil Mechanics and Foundation Engineering, Moscow*, vol. 3, pp 111-159.

Bjerrum L. & Landva A. (1966). Direct simple shear tests on a Norwegian quick clay. *Géotechnique*, vol. 16, pp 1-20.

Bjerrum L. & Simons N.E. (1960). Comparison of shear strength characteristics of normally consolidated clays. *Proc. ASCE Conf. on Shear Strength of Cohesive Soils, Boulder, Colorado*, pp 711-726.

Black D.K. & Lee K.L. (1973). Saturating laboratory samples by back pressure. *ASCE Journal of the Soil Mechanics and Foundation Division*, vol. 99, SM1, pp 75-93.

Bolton M. (1979). *A guide to soil mechanics.* Macmillan, London, 439 pp.

Bolton M. (1986). The strength and dilatancy of sands. *Géotechnique*, vol. 36:1.

Bowles J.E. (1970). *Engineering properties of soils and their measurement.* McGraw-Hill, New York, 187 pp.

Britto A. & Gunn M.J. (1987). *Critical state soil mechanics via finite element.* Ellis Horwood, Chichester, 488 pp.

Bromhead E.N. (1979). A simple ring shear apparatus. *Ground Engineering*, vol. 12:5.

Bromhead E.N. (1986). *The stability of slopes.* Surrey University Press.

Bromhead E.N. & Curtis R.D. (1983). Comparison of alternative methods of measuring residual strength of London Clay. *Ground Engineering*, vol. 16:4.

Burland J.B. (1967). Deformation of soft clays. Phd thesis, Cambridge University.

Cadling L. & Odenstad S. (1950). The vaneborer: an apparatus for determining the shear strength of soils directly in the ground. *Proc. Royal Swedish Geotechnical Institute*, no. 2, 87 pp.

Campanella R.G., Robertson P.K. & Gillespie D. (1983). Cone penetration in deltaic soils. *Canadian Geotechnical Journal*, vol. 20:1, pp 23-35.

Carrier W.D. (1985). Consolidation parameters derived from index tests. *Géotechnique*, vol. 35:2, pp 211-213.

Carrier W.D. & Beckman J.F. (1984). Correlations between index tests and the properties of remoulded clays. *Géotechnique*, vol. 34:2, pp 211-228.

Casagrande A. (1936). The determination of the preconsolidation load and its practical significance. *1st ICSMFE, Cambridge*, vol. 3, p. 60.

Casagrande A. (1948). Classification and identification of soils. *Transactions ASCE*, vol. 113, pp 901-930.

Casagrande L. (1966). Construction of embankments across peaty soils. *Journal of the Boston Society of Civil Engineers*, vol. 53:3, pp 272-317.

Cedergren H.C. (1977). *Seepage, drainage and flownets.* Wiley Interscience, 534 pp.

Chandler R.J. (1987). The in situ measurement of the undrained shear strength of clays using the field vane. State-of-the-art paper presented at the ASTM Symp. on Laboratory and Field Vane Shear Strength Testing of Soils, STP 1014, Tampa, Fla.

Charles J.A. & Soares M.M. (1984). Stability of compacted rockfill slopes. *Géotechnique*, vol. 34:1, pp 61-70.

Costa-Filho L.M., Aragão C.J.G. & Velloso P.P.C. (1985). Características geotécnicas de alguns depósitos de argila mole na área do Rio de Janeiro. *Solos e Rochas*, vol. 8:1, pp 3-13.

Costa-Filho L.M., Werneck M.L.G. & Collet H.B. (1977). The undrained strength of a very soft clay. *Proc. 9th ICSMFE, Int. Conf. on Soil Mechanics and Foundation Engineering, Tokyo*, vol. 1, pp 69-72.

Coutinho R. & Ortigao J.A.R. (1990). O desempenho da instrumentação de um aterro sobre solo mole. *Singeo 90 – Simp. Sobre Instrumentação Geotécnica de Campo, Rio de Janeiro*, pp 95-105.

Cozzolino. V.M. (1961). Statistical forecasting of compression index. *Proc. 5th ICSMFE, Paris*.

Cruz P.T. (1967). Propriedades de engenharia de solos residuais compactados da região Centro-Sul do Brasil. Escola Politécnica, Universidade de São Paulo, 191 pp.

Daniel D. (1989). In situ hydraulic conductivity tests for compacted clays. *ASCE Journal of Geotechnical Engineering*, vol. 115:9, Sept. 1989, pp 1205-1226.

Danziger F.A.B. (1990). Desenvolvimento de equipamento para realização de ensaio de piezocone: aplicação a argilas moles. DSc thesis, Universidade Federal do Rio de Janeiro, 466 pp.

Daramola O. (1980). On estimating K_0 for overconsolidated granular soils. *Géotechnique*, vol. 30:3, pp 310-313.

Datta M., Gulhati S.K. & Rao G.V. (1980). An appraisal of the existing practice of determining the axial load of deep penetration piles in calcareous sands. *OTC Offshore Technology Conference, Houston, Texas*, paper OTC 3867, pp 119-130.

Desai C. & Christian J.T. (1977). *Numerical methods in geotechnical engineering*. McGraw-Hill, New York.

Desai C. & Siriwardane H.J. (1984). *Constitutive laws for engineering materials with emphasis on geologic materials*. Prentice Hall, New Jersey, 468 pp.

Dias R.D. & Gehling W.Y.Y. (1983). Considerações sobre solos tropicais. Caderno. Técnico, CPGEC Curso de Pós-Graduação em Engenharia Civil, Universidade Federal do Rio Grande do Sul, Porto Alegre, 59 pp.

ENR Enginnering News Record (1977). Teton dam failure is blamed on USBR design deficiencies, vol. 13, pp 8-9.

Fadum R.E. (1948). Influence values for estimating stresses in elastic foundations. *Proc. 2nd ICSMFE, Rotterdam*, vol. 3, pp 77-84.

Fell R. & Jeffery R. (1987). Determination of drained shear strength for slope stability analysis. In: Walker B.F. & Fell R. (eds), *Soil slope instability and stabilization*. Balkema, Rotterdam, pp 53-70.

Ferreira R.C. & Monteiro L.B. (1985). Identification and evaluation of colluvial soils that occur in the São Paulo State. *1st Int. Conf. on Geomechanics in Tropical Lateritic and Saprolitic Soils, Brasília*, vol. 1, pp 269-280.

Flodin N. & Broms B.B. (1981). *History of civil engineering of soft clays: Soft clay engineering*. Elsevier, Amsterdam, pp 27-156.

Franciss F.O. (1980). *Hidráulica dos meios permeáveis*. Interciência, Rio de Janeiro.

Gallagher K.A. (1983). Personal communication.

Gerscovich D.M., Costa-Filho L.M. & Bressani L.A. (1986). Propriedades geotécnicas da camada ressecada de um depósito de argila mole da Baixada Fluminense. *Anais do 8° COBRAMSEF Congr. Bras. de Mecânica dos Solos e Fundações, Porto Alegre*, vol. 2, pp 289-300.

Giroud J.P. (1975). *Tassement et stabilité des fondations superficielles*. Presse Universitaires de Grenoble, 243 pp.

Gonzalez M.D., Dias R.D. & Roisenberg A. (1981). Contribuição ao estudo do comportamento de solos argilosos da região de Manaus. *Simp. Bras. de Solos Tropicais em Engenharia, Universidade Federal do Rio de Janeiro*, vol. 1, pp 165-178.

Greig J.W., Campanella R.G. & Robertson P.K. (1987). Comparison of field vane results with other in situ test results. *ASTM Symp. on Laboratory and Field Vane Shear Strength Testing, Tampa, Florida*, STP 1014, pp 247-263.

Harr M.E. (1962). *Groundwater and seepage*. McGraw-Hill, New York, 315 pp.

Harr M.E. (1966). *Foundations of theoretical soil mechanics*. McGraw-Hill, New York, 381 pp.

Head K.H. (1986). *Manual of laboratory soil testing*. Pentech Press, London, vol. 1, 2 and 3.

Henkel D.J. (1960). The shear strength of saturated remoulded clays. *Proc. ASCE Conf. on Shear Strength of Cohesive Soils, Boulder, Colorado*, pp 533-554.

Herrero O.R. (1980). Universal compression index equation. *ASCE Journal of Geotechnical Engineering*, vol. 106, GT11, pp 1179-1200.

Höeg K., Andersland O.B. & Rolfsen E.N. (1969). Undrained behaviour of quick clay under load test at Åsrum. *Géotechnique*, vol. 19:1, pp 101-115.

Hoek E. (1983). Strength of jointed rock masses, Rankine Lecture. *Géotechnique*, vol. 33:3, pp 187-223.

Holl D.L. (1940). Stress transmission in earths. *Proc. Highway Research Board*, vol. 20, pp 709-721.

Holtz R.D. & Kovacs W.D. (1981). *An introduction to geotechnical engineering*. Prentice-Hall, New Jersey, 733 pp.

Houlsby G.T. & Teh C.I. (1988). Analysis of the piezocone tests in clay. In: De Ruiter J. (ed.), *Penetration testing 1988, Proc. 1st ISOPT, Orlando*. Balkema, Rotterdam, vol. 2, pp 777-783.

Hunt R.E. (1984). *Geotechnical engineering investigation manual*. McGraw-Hill, New York, 984 pp.

Hvorslev M.J. (1951). Time lag and soil permeability in ground water observation. US Army Engineers Waterways Experiment Station, Bulletin no. 36, Vicksburg, Miss., 50 pp.

Jaky J. (1944). The coefficient of earth pressure at rest. *Journal of the Society of Hungarian Architects and Engineers*, pp 355-358.

Janbu N. (1963). Soil compressibility as determined by oedometer and triaxial tests. *ECSMFE, Wiesbaden*, vol. 1, p 1925, and vol. 2, p 1721.

Kenney T.C. (1959). Discussion. *Proc. ASCE*, vol. 85, no. SM3, pp 67-79.

Kraft L.M. (1982). Effective stress capacity model for piles in clay. *ASCE Journal of Geotechnical Engineering*, vol. 108:11, pp 1387-1403.

Lacerda W.A. (1985). Compressibility properties of lateritic and saprolitic soils. Progress Report, Committee on Tropical Soils of the ISSMFE. *1st Int. Conf. on Tropical Lateritic and Saprolitic Soils, Brasília*, pp 37-65.

Ladd C.C. (1971). Strength parameters and stress-strain behaviour of saturated clays. MIT Research Report R71-23, Soils Publication no. 278.

Ladd C.C. (1973). Estimating settlement of structures supported on cohesive soils. MIT Soils Publication no. 272, 99 pp.

Ladd C.C. & Foott R. (1974). New design procedure for stability of soft clays. *ASCE Journal of the Geotechnical Engineering Division*, vol. 100:7, pp 763-786.

Ladd C.C., Foott R., Ishihara K., Schlosser F. & Poulos H.G. (1977). Stress deformation and strength characteristics. *Proc. 9th ICSMFE Int. Conf. on Soil Mechanics and Foundation Engineering, Tokyo*, vol. 2, pp 421-494.

Lambe T.W. (1951). *Soil testing for engineers*. Wiley, New York, 165 pp.

Lambe T.W. & Whitman R.V. (1979). *Soil mechanics – SI version*. Wiley, New York, 553 pp.

Lee K.L. (1965). Triaxial compressive strength of saturated sands under seismic loading conditions. Phd thesis, University of California, Berkeley, 521 pp.

Leonards G.A. (1962). Engineering properties of soils. In: Leonards G.A. (ed.), *Foundation engineering*. McGraw-Hill, pp 66-240.

Leroueil S., Magnan J.P. & Tavenas F. (1985). *Remblais sur argiles molles*. Technique et Documentation Lavoisier, Paris, 342 pp.

Leroueil S., Tavenas F., Mieussens C. & Peignaud M. (1978). Construction pore pressures in clay foundations under embankments, Part II: Generalized behaviour. *Canadian Geotechnical Journal*, vol. 15:1, pp 66-82.

Leroueil S., Tavenas F., Trak B., La Rochelle P. & Roy M. (1978). Construction pore pressures in clay foundations under embankments, Part I: The St Albans test fills. *Canadian Geotechnical Journal*, vol. 15:1, pp 54-65.

Lopes F.R. (1985). Lateral resistance of piles in clay and possible effect of loading rate. *Symp. on Theory and Practice in Deep Foundations, Porto Alegre, RS*, vol. 1, pp 53-68.

Lupini J.F., Skinner A.E. & Vaughan P.R. (1981). The drained residual strength of cohesive soils. *Géotechnique*, vol. 31:2, pp 181-213.

Massad F., Rocha J.L.R. & Yassuda A.Y. (1981). Algumas características geotécnicas de solos da formação Guabirotuba, Paraná. *Simp. Bras. de Solos Tropicais em Engenharia, Universidade Federal do Rio de Janeiro*, vol. 1, pp 706-723.

Mayne P.W. (1980). Cam-clay predictions of undrained strength. *ASCE Journal of the Geotechnical Engineering Division*, vol. 106, GT11, pp 1219-1242.

Mayne P.W. & Kulhawy F.H. (1982). K_0-*OCR* relationship in soil. *ASCE Journal of Geotechnical Engineering*, vol. 108, GT6, pp 851-872.

Means R.E. & Parcher J.V. (1965). *Physical properties of soils*. Prentice Hall, 464 pp.

Meigh A.C. (1987). *Cone penetration testing: methods and interpretation*. Butterworths, London, 141 pp.

Mello V.F.B. de (1977). Reflections on design decisions of practical significance to embankment dams, 17th Rankine Lecture. *Géotechnique*, vol. 27:3, pp 281-354.

Mesri G. (1975). Discussion on new design procedure of stability of soft clays. *ASCE Journal of the Geotechnical Engineering Division*, vol. 101:4, pp 409-412.

Mesri G. & Cepeda-Diaz A.F. (1986). Residual shear strength of clays and shales. *Géotechnique*, vol. 36:2, pp 269-274.

Milititsky J. (1986). Fundações, Relatório Geral. *8° COBRAMSEF Congresso Bras de Mec dos Solos e Fundações, Porto Alegre*, vol. 7, pp 191-260.

Milititsky J. & Dias R.D. (1985). Fundações diretas sobre solos tropicais. Caderno. Técnico, CPGEC Curso de Pós-Graduação em Engenharia Civil, Universidade Federal do Rio Grande do Sul, Porto Alegre.

Miranda A.N. (1988). Behaviour of small earth dams during initial filling. Phd thesis, Colorado State University, Fort Collins, CO.

Mitchell J.K. (1976). *Fundamentals of soil behaviour*. Wiley, 422 pp.

Mitchell R.J. (1983). Earth structures engineering, Allen & Unwin, Ontario, 265 pp.

Mitchell J.K. & Houston W.N. (1969). Causes of clay sensitivity. *ASCE Journal of Soils Mechanics and Foundation Engineering Division*, vol. 95, SM3, pp 845-871.

Mori R.T., Freitas M.S. & Bertolucci J.C. (1974). Estudos de compressibilidade de solos residuais e transportados de basalto. *Anais 8° COBRAMSEF, São Paulo*, vol. 2, pp 193-206.

Nadarajah V. (1973). Stress-strain properties of lightly overconsolidated clays. Phd thesis, Cambridge University.

Navfac DM7 (1971). Design manual soil mechanics, foundations and earth structures. Naval Facilities Engineering Command, Washington DC.

Nunes A.J.C. (1971). Personal communication.

Nunes A.J.C. (1978). Fundações em terrenos expansivos. *1° Sem. Regional de Mecânica dos Solos e Engenharia de Fundações, Salvador*, vol. 2.

Nuñez E. & Micucci C.A. (1985). Engineering parameters in residual soils. *1st Int. Conf. on Tropical Lateritic and Saprolitic Soils, Brasília*, vol. 1, pp 383-396.

Ortigao J.A.R. (1975). Contribuição ao estudo das propriedades geotécnicas de um depósito de argila mole da Baixada Fluminense. MSc thesis, Universidade Federal do Rio de Janeiro, 94 pp.

Ortigao J.A.R. (1978). Efeito do pré-adensamento e da consolidação anisotrópica em algumas propriedades da argila mole da Baixada Fluminense. *Anais do 6° COBRAMSEF Congr. Bras. de Mecânica dos Solos e Fundações, Rio de Janeiro*, vol. 1, pp 243-259.

Ortigao J.A.R. (1988). Experiência com ensaios de palheta em terra e no. mar. *Simp. sobre Novos Conceitos em Ensaios de Campo e Laboratório, Universidade Federal do Rio de Janeiro, Rio de Janeiro*, vol. 3, pp 157-180.

Ortigao J.A.R. & Almeida M.S.S. (1988). Stability and deformation of embankments on soft clay. In: Chereminisoff P.N., Chereminisoff N.P. & Cheng S.L. (eds), *Handbook of civil engineering practice*. Technomics, New Jersey, vol. III, Geotechnics, pp 267-336.

Ortigao J.A.R. & Collet H.B. (1987). Errors caused by friction in field vane testing. *ASTM Symp. on Laboratory and Field Vane Shear Strength Testing, Tampa, Florida*, STP 1014, pp 104-116.

Ortigao J.A.R. & Macedo P. (1993). Large settlements due to tunneling in porous clay. *Tunnels et ouvrages souterrains, AFTES Association Française des Tunnels et Ouvrages Souterrains*, no. 119, Sept.-Oct. 93, pp 245-250, Paris.

Ortigao J.A.R. & Randolph M.F. (1983). Creep effects on tension piles for the design of buoyant offshore structures. *Int. Symp. on Offshore Engineering, Brazil Offshore '83. Rio de Janeiro*, Pentech Press, London, pp 478-498.

Ortigao J.A.R. & Sayão A.S.F.J. (1994). Settlement characteristics of a soft clay. ASCE Specialty Conference: Settlement 94, Texas A & M University, Austin, TX, June 1994. vol. 2, pp 1415-1424.

Ortigao J.A.R., Capellão S.L.F. & Delamonica L. (1985). Marine site investigation and assessment of calcareous sand behaviour at the Campos basin, Brazil. *Int. Symp. on Offshore Engineering, Brazil Offshore '85, Rio de Janeiro*. Pentech Press, London, pp 238-255.

Ortigao J.A.R., Cortes H.V.M. & Medeiros C.J. (1985). Performance observations for the assessment of the behaviour of offshore piles in calcareous soils: a review. *Symp. on Theory and Practice in Deep Foundations, Porto Alegre, RS*, vol. 1, pp 341-357.

Ortigao J.A.R., Coutinho R.Q. & Sant'Anna L.A.M. (1987). Discussion on embankment failures on soft clay in Brazil. *Proc. Int. Symp. on Geotechnical Engineering of Soft Soils, Mexico*. Submitted for publication, vol. 2.

Ortigao J.A.R., Werneck M.L.G. & Lacerda W.A. (1983). Embankment failure on Rio de Janeiro clay. *ASCE Journal of Geotechnical Engineering*, vol. 109:11, pp 1460-1479.

Ortigao J.A.R., Capellão S.L.F., Morrison M. & Delamonica L. (1986). In situ testing of calcareous sand, Campos basin. *ASCE Specialty Conf. on Use of In Situ Testing in Geotechnical Engineering, In Situ '86, Virginia Tech, Blacksburg*, Geotechnical Publication no. 6, pp 887-899.

Pacheco E.B. (1978). Estudo da resistência ao cisalhamento e densidade relativa das areias e correlações com o SPT. MSc thesis, Federal University of Rio de Janeiro, 119 pp.

Palmeira E.M. (1987). The study of soil reinforcement interaction by means of large scale laboratory tests. Phd thesis, University of Oxford.

Pavlovsky N.N. (1956). Collection of papers. Akad. Leningrad, USSR. Referenced by Harr (1962).

Perrin J. (1973). Comportement des sols toubeux et synthese des resultats. Remblais sur Sols Compressibles, Bulletin des Liaison des Laboratoire des Ponts et Chaussées, Special T, pp 208-217.

Polido U. & Castelo R.R. (1985). Geotechnical parameters of a lateritic soil of the Barreiras formation in Espírito Santo. *1st Int. Conf. on Tropical Lateritic and Saprolitic Soils, Brasília*, vol. 1, pp 403-416.

Polubarinova-Kochina P.Y. (1962). *Theory of ground water movement*. Princeton University Press, 613 pp.

Poulos H. & Davis E.H. (1974). *Elastic solutions for soil and rock mechanics*. Wiley.

Prakash S. (1981). *Soil dynamics*. McGraw-Hill, New York, 426 pp.

Reichardt K. (1985). *Processos de transferência no. sistema solo-planta-atmosfera*. Campinas, Fundação Cargill, 466 pp.

Roberts J.E. (1964). Sand compression as a factor in oil field subsidence. ScD thesis, MIT.

Robertson P. & Campanella R.G. (1989). Guidelines for geotechnical design using CPT and CPTU. Soil Mechanics Series no. 120, Department of Civil Engineering, The University of British Columbia, Vancouver, 193 pp.

Rocha P. & Alencar J.A. (1985). Piezocone tests in the Rio de Janeiro clay. *Proc. 11th ICSMFE, San Francisco*. Balkema, Rotterdam.

Roscoe K.H. (1970). The influence of strains in soil mechanics, The 10th Rankine Lecture. *Géotechnique*, vol. 20, pp 129-170.

Rowe P.W. (1961). The stress-dilatancy relation for static equilibrium of an assembly of particles in contact. *Proc. of The Royal Society, London*, Series A, vol. 269, pp. 500-527.

Rowe P.W. (1963). Stress-dilatancy, earth pressures and slopes. *ASCE Journal of the SMFD*, vol. 89, SM3, pp 37-61.

Rushton K.R. & Redshaw S.C. (1978). *Seepage and groundwater flow*. Wiley, London, 339 pp.

Sayão A.S.F & Vaid Y.P. (1988). Stress path testing in a hollow cylinder torsional device. *Simp. Novos Conceitos em Ensaios de Campo e Laboratorio em Geotecnia, Universidade Federal do Rio de Janeiro*, vol. 1, pp 197-120.

Schofield A.N. & Wroth C.P. (1968). *Critical state soil mechanics*. McGraw-Hill, London, 310 pp.

Silva P.F. (1970). Uma construção gráfica para a determinação da pressão de pré-adensamento de uma amostra de solo. *COBRAMSEF, Rio de Janeiro*, vol. 2, pp 219-223.

Simões P.R. & Costa-Filho L.M. (1981). Características mineralógicas, químicas e geotécnicas de solos expansivos do recôncavo baiano. *Simp. Bras. de Solos Tropicais em Engenharia, Universidade Federal do Rio de Janeiro*, vol. 1, pp 569-588.

Skempton A.W. (1948). The $\phi_u = 0$ analysis of stability and its theoretical basis. *Proc. 2nd ICSMFE Int. Symp. on Soil Mechanics and Foundation Engineering, Rotterdam*, vol. 1, pp 72-78.

Skempton A.W. (1953). The colloidal activity of clays. *Proc. 3rd ICSMFE*, vol. 1, pp 57-61.

Skempton A.W. (1954). The pore-pressure coefficients A and B. *Géotechnique*, vol. 4:4, pp 143-147.

Skempton A.W. & Northey R.D. (1952). The sensitivity of clays. *Géotechnique*, vol. 3:1, pp 30-53.

Sowers G.E. (1963). Engineering properties of residual soils derived fom igneous and metamorphic rocks. *Proc. 2nd Pan. Conf. on Soil Mechanics and Foundation Engineering, Brazil*, vol. 1, pp 39-62.

Sully J.P. & Campanella R.G. (1989). Lateral stress measurements in a glaciomarine silty clay. *Proc. of the 25th Conf. on Quaternary Engineering Geology, Edinburgh*. Engineering Geology Group, Geological Society, London.

Tavenas F., Jean P., Leblond P. & Leroueil S. (1983). The permeability of natural soft clays, Part II: Permeability characteristics. *Canadian Geotechnical Journal*, vol. 20:4, pp 645-660.

Tavenas F., Tremblay M., Larouche G. & Leroueil S. (1986). In situ measurement of permeability in soft clays. *ASCE Symp. on Use of In Situ Tests in Geotechnical Engineering, In Situ '86*. Geotechnical Publication no. 6, pp 1034-1048.

Taylor D.W. (1948). *Fundamentals of soil mechanics*. Wiley, 700 pp.

Teixeira A.H. (1988). Capacidade de carga de estacas pré-moldadas de concreto nos sedimentos quaternários da Baixada Santista. *Sidequa Simp. sobre Depósitos Quanternários das Baixadas Brasileiras, Rio de Janeiro*, vol. 2, pp 5.1-5.25.

Terzaghi K. & Peck R.B. (1967). *Soil mechanics in engineering practice*. Wiley, New York.

Thomas S.D. (1986). Various techniques for the evaluation of the coefficient of consolidation from a piezocone dissipation test. Research Report SM064/86, Oxford University, Oxford.

Timoshenko S. & Goodier J.N. (1951). *Theory of elasticity*. McGraw-Hill, New York.

USCS (1970). Laboratory soils testing. Engineer Manual EM 1110-2-1906, US Corps of Engineers, Washington DC.

Vargas M. (1953). Some engineering properties of residual soils occurring in southern Brazil. *Proc. 3rd ICSMFE, Zürich*.

Vargas M. (1973). Structurally unstable soils in southern Brazil. *Proc. 8th ICSMFE, Moscow*.

Vargas M. (1977). *Introdução à mecânica dos solos*. McGraw-Hill, São Paulo, 509 pp.

Verruijt A. (1982). *Theory of groundwater flow*. MacMillan, London, 141 pp.

Vesic A.S. & Clough G.W. (1968). Behaviour of granular materials under high stresses. *ASCE Journal of Soil Mechanics and Foundation Division*, vol. 94, SM3, pp 661-688.

Vickers B. (1978). *Laboratory work in civil engineering soil mechanics*. Granada, UK, 148 pp, Engineering, vol. 16:4.

Vilar O.M., Rodrigues J.E. & Nogueira J.B. (1981). Solos colapsíveis: um problema para a engenharia de solos tropicais. *Simp. Bras. de Solos Tropicais em Engenharia, Universidade Federal do Rio de Janeiro*, vol. 1, pp 209-224.

Wong K.S. & Duncan J.M. (1974). Hyperbolic stress-strain parameters for nonlinear finite element analyses of stresses and movements in soil masses. Report no. TE-74-3, University of California, Berkeley, 90 pp.

Subject index

297

PUBLISHER'S NOTE

For your convenience, please use the enclosed postcard to order the computer programs mentioned in this book. If this postcard is no longer in the book, kindly photocopy the blank order form reproduced underneath, complete it, and mail it to the publisher together with your payment. You have permission from the publisher to photocopy this page.

Milton Keynes UK
Ingram Content Group UK Ltd.
UKHW051932141024
449569UK00027B/1461